*INTRODUCTION TO*
# STATISTICAL PHYSICS

# INTRODUCTION TO
# STATISTICAL PHYSICS

## KERSON HUANG

*Professor of Physics Emeritus*
*Massachusetts Institute of Technology*
*Cambridge, Massachusetts*
*USA*

## CRC PRESS

Boca Raton   London   New York   Washington, D.C.

## Library of Congress Cataloging-in-Publication Data

Huang, Kerson, 1928-
   Introduction to statistical physics / Kerson Huang.
      p. cm. .
   Includes bibliographical references and index.
   ISBN 0-7484-0942-4  (alk. paper)
   1. Statistical physics. I. Title.
QC174.8 .H82   2001
530.15′95—dc21                                                    2001023960

**Visit the CRC Press Web site at www.crcpress.com**

© 2001 by Kerson Huang

No claim to original U.S. Government works
International Standard Book Number 0-7484-0942-4
Library of Congress Catalog Number 1002023960
Printed in the United States of America        3  4  5  6  7  8  9  0
Printed on acid-free paper

To the memory of Herman Feshbach (1917–2000)

# Contents

# Preface

This book is based on the lecture notes for a one-semester course on statistical physics at M.I.T., which I have taught to advanced undergraduates. They were expected to have taken introductory courses in classical and quantum physics, including thermodynamics. The object of this book is to guide the reader quickly but critically through a statistical view of the physical world.

The tour is divided into stages. First we introduce thermodynamics as a phenomenological theory of matter, and emphasize the power and beauty of such an approach. We then show how all this can be derived from the laws governing atoms, with the help of the statistical approach. This is done using the ideal gas, both classically and quantum mechanically. We show what a wide range of physical phenomena can be explained by the properties of the Bose gas and the Fermi gas. Only then do we launch into the formal methods of statistical mechanics, the canonical ensemble and the grand canonical ensemble.

In the last stage we return to phenomenology, using insight gained from the microscopic level. The order parameter is introduced to describe broken symmetry, as manifested in superfluidity and superconductivity. We spend quite some time on noise, beginning with Einstein's description of Brownian motion, through stochastic processes and time-series analysis, and ending with the Monte-Carlo method.

There is more material in this book than one could comfortably cover in one semester, perhaps. As a rough division, twelve chapters in this book are on "general topics" and six on "special applications". A course might cover all of the general topics, with choices of special applications that vary from year to year. Among the general topics are Chapters 1–3 on thermodynamics, Chapters 5, 6 and 8 on the kinetic theory of the ideal gas, Chapters 9 and 10 on the Bose and Fermi gases, Chapters 12 and 13 on formal statistical mechanics, and Chapters 17 and 18 on stochastic processes. That leaves six chapters on such things as transport phenomena, Bose–Einstein condensation, Josephson junction, and Brownian motion.

The emphasis of this book is on physical understanding rather than calculational techniques, although Dirac would say that to understand is to know how to calculate. I have included quite a few problems at the end of each chapter for the latter purpose.

I would like to thank the M.I.T. students who had endured 8.08, and Aleksey Lomakin and Patrick Lee, with whom I have enjoyed teaching that course, which led to the writing of this book.

Kerson Huang
Boston, Massachusetts, 2001

# Chapter 1

# The macroscopic view

## 1.1 Thermodynamics

The world puts on different faces for observers using different scales in the measurement of space and time. The everyday, macroscopic world, as perceived on the scale of meters and seconds, looks very different from that of the atomic microscopic world, which is seen on scales smaller by some ten orders of magnitude. Different still is the submicroscopic realm of quarks, which is revealed only when the scale shrinks further by another ten orders of magnitude. With an expanding scale in the opposite direction, one enters the regime of astronomy, and ultimately cosmology. These different pictures arise from different ways of organizing data, while the basic laws of physics remain the same.

The physical laws at the smallest accessible length scale are the most "fundamental", but they are of little use on a larger scale, where we must deal with different physical variables. A complete knowledge of quarks tells us nothing about the structure of nuclei, unless we can define nuclear variables in terms of quarks, and obtain their equations of motion from those for quarks. Needless to say, we are unable to do this in detail, although we can see how this could be done in principle. Therefore, each regime of scales has to be described phenomenologically, in terms of variables and laws observable in that regime. For example, an atom is described in terms of electrons and nuclei without reference to quarks. The subnuclear world enters the equations only implicitly, through such parameters as the mass ratio of electron to proton, which we take from experiments in the atomic regime. Similarly, in the macroscopic domain, where atoms are too small to be visible, we describe matter in terms of phenomenological variables such as pressure and temperature. The atomic structure of matter enters the picture implicitly, in terms of properties such as density and heat capacity, which can be measured by macroscopic instruments.

From experience, we know that a macroscopic body generally settles down, or "relaxes" to a stationary state after a short time. We call this a state of *thermal equilibrium*. When the external condition is changed, the existing equilibrium state will change, and, after a relatively short relaxation time, settles down to another equilibrium state. Thus, a macroscopic body spends most of the time in some state of equilibrium, punctuated by almost sudden transitions. In our study of macroscopic phenomena, we divide the subject roughly under the following headings:

- *Thermodynamics* is a phenomenological theory of equilibrium states and transitions among them.
- *Statistical mechanics* is concerned with deducing the thermodynamic properties of a macroscopic system from its microscopic structure.
- *Kinetic theory* aims at a microscopic description of the transition process between equilibrium states.

This book is concerned mainly with statistical methods, which provide a bridge between the microscopic and the macroscopic world. We begin our study with thermodynamics, because it is a highly successful phenomenological theory, which identifies the correct macroscopic variables to use, and serves as a guide post for statistical mechanics.

## 1.2 Thermodynamic variables

As a rule, properties of a macroscopic system can be classified as either extensive or intensive:

- *Extensive* quantities are proportional to the amount of matter present.
- *Intensive* quantities are independent of the amount of matter present.

Generally, there are only these two categories, because we can neglect surface effects. A macroscopic body is typically of size $L \sim 1$ m, while the range of atomic forces is of order $r_0 \sim 10^{-10}$ m. The macroscopic nature is expressed by the ratio $L/r_0 \sim 10^{10}$. The surface-to-volume ratio, rendered dimensionless in terms of the range of atomic forces, is of order $r_0/L \sim 10^{-10}$. The extensive property expresses the "saturation property" of atomic forces, i.e., an atom can "feel" only as far as the range of the force. The intensive property means that atoms in the interior of the body do not feel the presence of the surface.

Exceptions arise when any of the following conditions prevail:

- The system is small.
- There is a non-uniform external potential.
- There are long-ranged interparticle forces, such as the Coulomb repulsion between charges, and the gravitational attraction between mass elements.
- The geometry is such that the surface is important.

These exceptions occur in important physical systems. For example, the volume of a star is a non-linear function of its mass, due to the long-ranged gravitational interaction. We illustrated the different cases in Fig. 1.1.

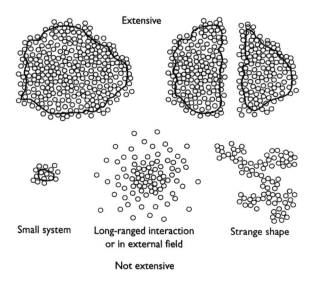

Extensive

Not extensive

Small system     Long-ranged interaction     Strange shape
or in external field

*Figure 1.1* A body has extensive properties if surface effects can be neglected, so that the energy is proportional to the number of particles. In these pictures, the surface layer is indicated by a heavy line.

## 1.3   Thermodynamic limit

We consider a material body consisting of $N$ atoms in volume $V$, with no non-uniform external potential, in the idealized limit

$$N \rightarrow \infty, \qquad V \rightarrow \infty,$$
$$\frac{N}{V} = \text{fixed number.} \tag{1.1}$$

This is called the *thermodynamic limit*, in which the system is translationally invariant.

The *thermodynamic state* is specified by a number of thermodynamic variables, which are assumed to be either extensive (proportional to $N$), or intensive (independent of $N$). We consider a generic system described by the three variables $P$, $V$ and $T$, denoting, respectively, the pressure, volume and the temperature:

- The pressure $P$, an intensive quantity, is the force per unit area that the body exerts on a wall, which can be that of the container of the system, or it may be one side of an imaginary surface inside the body. Under equilibrium conditions in the absence of external potentials, the pressure must be uniform throughout the body. It is convenient to measure pressure in terms of the atmospheric pressure at sea level:

$$1 \, \mathrm{atm} = 1.013 \times 10^6 \, \mathrm{dyne \, cm^{-2}} = 1.103 \times 10^5 \, \mathrm{N \, m^{-2}}. \qquad (1.2)$$

- The volume $V$ measures the spatial extent of the body, and is an extensive quantity. A solid body maintains its own characteristic density $N/V$. A gas, however, fills the entire volume of its container.
- The temperature $T$, an intensive quantity, is measured by some thermometer. It is an indicator of thermal equilibrium. Two bodies in contact with each other in equilibrium must have the same temperature. Since the two bodies in question can be different parts of the same body, the temperature of a body in equilibrium must be uniform. The temperature is also a measure of the energy content of a body, but the notion of energy has yet to be defined, through the first law of thermodynamics.

In the thermodynamic limit we must use only intensive quantities, mathematically speaking. Instead of $V$ we should use the specific volume $v = V/N$, or the density $n = N/V$. However, it is convenient to regard $V$ as a large but finite number, for this corresponds to the everyday experience of seeing the volume of a macroscopic body expand or contract, while the number of atoms is fixed.

There are systems requiring other variables in addition to, or in place of, $P$, $V$ and $T$. Common examples are the magnetic field and magnetization for a magnetic substance, the strain and stress in elastic solids, or the surface area and the surface tension.

## 1.4   Thermodynamic transformations

When a body is in thermal equilibrium, the thermodynamic variables are not independent of one another, but are constrained by an *equation of state* of the form

$$f(P, V, T) = 0, \qquad (1.3)$$

where the function $f$ is characteristic of the substance. This leaves two independent variables out of the original three. Geometrically we can represent the equation of state by a surface in the state space spanned by $P, V, T$, as shown in Fig. 1.2. All equilibrium states must lie on this surface. We regard $f$ as a continuous differentiable function, except possibly at special points.

A change in the external condition will change the equilibrium state of a system. For example, application of external pressure will cause the volume of a body to decrease. Such a change is called a *thermodynamic transformation*. The initial and final states are equilibrium states. If the transformation proceeds sufficiently slowly, the system can be considered to remain in equilibrium. In such a case, we say that the transformation is *quasi-static*. This usually means that the transformation is *reversible*, in that the system will retrace the transformation in reverse when

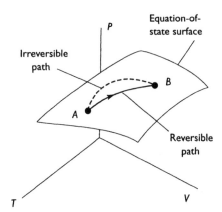

*Figure 1.2* The state space in thermodynamics.

the external change is reversed. A reversible transformation can be represented by a continuous path on the equation-of-state surface, as illustrated in Fig. 1.1.

An *irreversible* transformation, on the other hand, cannot be represented by a path on the equation-of-state space, as indicated by the dashed line in Fig. 1.1. In fact, we may not be able to represent it as a path in the state space at all. An example is the sudden removal of a wall in a container of a gas, so that the gas expands into a compartment that was originally vacuous. Although the initial and final states are equilibrium states, the intermediate states do not have uniform $P$, $V$, $T$, and hence cannot be represented as points in the state space.

In a reversible transformation, we can consider mathematically infinitesimal paths. The mechanical work done by the system over an infinitesimal path is represented by a differential:

$$dW = P\,dV. \tag{1.4}$$

Along a finite reversible path $A \to B$, the work done is given by

$$\Delta W = \int_A^B P\,dV \tag{1.5}$$

which depends on the path connecting $A$ to $B$. This is the area underneath the path in a $P$–$V$ diagram. When the path is a closed cycle, the work done in one cycle is the area enclosed, as shown in Fig. 1.3. The work done along an irreversible path is generally not $\int P\,dV$. For example, in the free expansion of a gas into a vacuum, the system does not perform work on any external agent, and so $\Delta W = 0$.

A uniquely thermodynamic process is heat transfer. From the atomic point of view, it represents a transfer of energy in the form of thermal agitation. In thermodynamics, we define heat phenomenologically, as that imparted by a heating

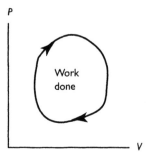

P

V

Figure 1.3 The work done in a closed cycle of transformations is represented by the area enclosed by the cycle in a P–V diagram.

element, such as a flame or a heating coil. An amount of heat $\Delta Q$ absorbed by a body causes a rise $\Delta T$ in its temperature given by

$$\Delta Q = C \, \Delta T, \tag{1.6}$$

where $C$ is the *heat capacity* of the substance. We imagine the limit in which $\Delta Q$ and $\Delta T$ become infinitesimal. The heat capacity is an extensive quantity. The intensive heat capacity per particle $(C/N)$, per mole $(C/n)$, or per unit volume $(C/V)$, are called *specific heats*.

The fact that heat is a form of energy was established experimentally, by observing that one can increase the temperature of a body by $\Delta T$ either by transferring heat to the body, or performing work on it. A practical unit for heat is the *calorie*, originally defined as the amount of heat that will raise the temperature of 1 g of water from 14.5°C to 15.5°C at sea level. In current usage, it is defined exactly in terms of the joule:

$$1 \, \text{cal} \equiv 4.184 \, \text{J}. \tag{1.7}$$

Another commonly used unit is the *British thermal unit* (Btu):

$$1 \, \text{Btu} \equiv 1,055 \, \text{J}. \tag{1.8}$$

The heat absorbed by a body depends on the path of the transformation, as is true of the mechanical work done by the body. We can speak of the amount of heat absorbed in a process; but the "heat of a body", like the "work of a body", is meaningless. Commonly encountered transformations are the following:

- $T = \text{constant}$  (Isothermal process)
- $P = \text{constant}$  (Isobaric process)
- $V = \text{constant}$  (Constant-volume process)
- $\Delta Q = 0$    (Adiabatic process).

We use a subscript to distinguish the different types of paths, as for example $C_V$ and $C_P$, representing, respectively, the heat capacity at constant volume and constant pressure. The heat capacity is only one of many thermodynamic coefficients that measure the response of the system to an external source. Other examples are

$$\kappa = -\frac{1}{V}\frac{\Delta V}{\Delta P} \quad \text{(Compressibility)},$$

$$\alpha = \frac{1}{V}\frac{\Delta V}{\Delta T} \quad \text{(Coefficient of thermal expansion)}. \tag{1.9}$$

These coefficients can be obtained from experimental measurements. In principle, they can be calculated from atomic properties using statistical mechanics.

## 1.5  Classical ideal gas

The simplest thermodynamic system is the classical ideal gas, which is a gas in the limit of low density and high temperature. The equation of state is given by the *ideal-gas law*

$$PV = NkT, \tag{1.10}$$

where $T$ is the *ideal gas temperature*, measured in kelvin (K), and

$$k = 1.381 \times 10^{-16} \, \text{erg K}^{-1} \quad \text{(Boltzmann's constant)}. \tag{1.11}$$

As we shall see, the second law of thermodynamics implies $T > 0$, and the lower bound is called the *absolute zero*. For this reason, $T$ is also called the *absolute temperature*. The heat capacity of a monatomic ideal gas at constant volume, $C_V$, has the value

$$C_V = \tfrac{3}{2}Nk. \tag{1.12}$$

These properties of the ideal gas were established experimentally, and can be derived theoretically in statistical mechanics.

Thermodynamics does not assume the existence of atoms. Instead of the number of atoms $N$, we can use the number of gram moles $n$, which is a chemical property of the substance. The two are related through

$$Nk = nR,$$

$$R = 8.314 \times 10^7 \, \text{erg K}^{-1} \quad \text{(Gas constant)}. \tag{1.13}$$

The ratio $R/k$ is Avogadro's number, the number of atoms per mole:

$$A_0 = \frac{R}{k} = 6.022 \times 10^{23}. \tag{1.14}$$

Indeed, the atomic picture emerged much later and gained acceptance only after a long historic struggle (see Chapter 16 on Brownian motion).

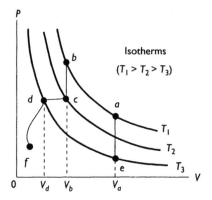

*Figure 1.4* Isotherms of an ideal gas, and various paths of transformations.

The equation of state can be represented graphically in a $P-V$ diagram, as shown in Fig. 1.4, which displays a family of curves at constant $T$ called *isotherms*. Indicated on this graph are the reversible paths corresponding to various types of transformations:

- *ab* is isothermal
- *bc* proceeds at constant volume
- *cd* is at constant pressure
- *de* is isothermal
- *abcdea* is a closed cycle
- *df* is non-isothermal.

To keep the temperature constant during an isothermal transformation, we keep the system in thermal contact with a body so large that its temperature is not noticeably affected by heat exchange with our system. We call such a body a *heat reservoir*, or *heat bath*.

## 1.6    First law of thermodynamics

The first law expresses the conservation of energy by including heat as a form of energy. It asserts that there exists *a* function of the state, *internal energy U*, whose change in any thermodynamic transformation is given by

$$\Delta U = \Delta Q - \Delta W, \tag{1.15}$$

that is, $\Delta U$ is independent of the path of the transformation, although $\Delta Q$ and $\Delta W$ are path-dependent. In a reversible infinitesimal transformation, the infinitesimal changes $dQ$ and $dW$ are not exact differentials, in the sense that they do not represent the changes of definite functions, but their difference

$$dU = dQ - dW \qquad (1.16)$$

is an exact differential.

## 1.7 Magnetic systems

For a magnetic substance, the thermodynamic variables are the intensive magnetic field $H$ and the extensive magnetization $M$, which are unidirectional and uniform in space. The magnetic field is generated by external real currents (and not induced currents), and the magnet work done by the system is given by

$$dW = -H \, dM. \qquad (1.17)$$

The first law takes the form

$$dU = dQ + H \, dM, \qquad (1.18)$$

which maps into that for a $PVT$ system under the correspondence $H \leftrightarrow -P$, $M \leftrightarrow V$.

For a paramagnetic substance, a magnetic field induces a magnetization density given by

$$\frac{M}{V} = \chi H. \qquad (1.19)$$

The magnetic susceptibility $\chi$ obeys Curie's law:

$$\chi = \frac{c_0}{T}. \qquad (1.20)$$

Thus, we have the equation of state

$$M = \frac{\kappa H}{T}, \qquad (1.21)$$

where $\kappa = c_0 V$.

An idealized uniform ferromagnetic system has a phase transition at a critical temperature $T_c$, and becomes a permanent magnet for $T < T_c$ in the absence of external field. However, there is no phase transition in the presence of an external field. The equation of state is illustrated graphically in the $M$–$T$ diagram of Fig. 1.5.

A real ferromagnet is not uniform, but breaks up into domains with different orientations of the magnetization. The configuration of the domains is such as to minimize the magnetic energy, by reducing the fringing fields that extend outside of the body. Each domain behaves according to the uniform case depicted in Fig. 1.5. At a fixed $T < T_c$, the domain walls move in response to a change in $H$, to either consolidate old domains or create new ones. This is a dissipative process, and leads to hysteresis, as shown in the $M$–$H$ plot of Fig. 1.6.

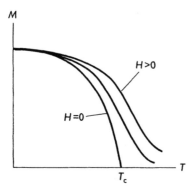

Figure 1.5 Equation of state of a uniform ferromagnet. A ferromagnetic transition occurs at a critical temperature at zero field.

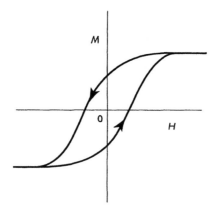

Figure 1.6 Hysteresis in a ferromagnet is due to the formation of ferromagnetic domains.

## Problems

**1.1**   Mr. Hillary heated 1 m³ of water from 20°C to 40°C, in order to take a bath before climbing Mt. Everest.

(a)  How much energy did he use? Give the answer in kwh.

(b)  Show that the energy is enough to lift Mr. Hillary's entire party of 14, of average weight 150 lb, from sea level to the top of Mt. Everest (elevation 29,000 ft).

**1.2**   Referring to Fig. 1.4, find the work done along the various paths on the closed cycle *ab*, *bc*, *cd*, *de*, *ea*, and give the total work done in the closed cycle. How much heat is supplied to the system in one cycle?

**1.3**   An ideal gas undergoes a reversible transformation along the path

$$\frac{V}{V_0} = \left(\frac{T}{T_0}\right)^b$$

where $V_0$, $T_0$ and $b$ are constants.

(a)  Find the coefficient of thermal expansion.

(b)  Calculate the work done by the gas when the temperature increases by $\Delta T$.

**1.4**   In a non-uniform gas, the equation of state is valid locally, as long as the density does not change too rapidly with position. Consider a column of ideal gas under gravity at constant temperature. Find the density as a function of height, by balancing the forces acting on a volume element.

**1.5**   Two solid bodies labeled 1 and 2 are in thermal contact with each other. The initial temperatures were $T_1$, $T_2$, with $T_1 > T_2$. The heat capacities are, respectively, $C_1$ and $C_2$. What is the final equilibrium temperature, if the bodies are completely isolated from the external world?

**1.6**   A paramagnetic substance is magnetized in an external magnetic field at constant temperature. How much work is require to attain a magnetization $M$?

**1.7**   A hysteresis curve (see Fig. 1.6) is given by the formula

$$M = M_0 \tanh\left(H \pm H_0\right),$$

where the $+$ sign refers to the upper branch, and the $-$ sign refers to the lower branch. The parameter $H_0$ is called the "coersive force". Show that the work done by the system in one hysteresis cycle is $-4M_0H_0$.

**1.8**   *The atomic nucleus*: The atomic nucleus contains typically fewer than 300 protons and neutrons. It is an example of a small system with both short- and long-ranged forces. The mass is given by the semi-empirical Weizsäcker formula (deShalit 1994):

$$M = Zm_p + Nm_n - a_1 A + a_2 A^{2/3} + a_3 \frac{Z^2}{A^{1/3}} + a_4 \frac{(Z-N)^2}{A} + \delta(A),$$

where $Z$ is the number of protons, $N$ is the number of neutrons, and $A = Z+N$. The masses of protons and neutrons are, respectively, $m_p$ and $m_n$, and $a_i (i = 1,\ldots,4)$ are numerical constants:

$$\text{volume energy: } a_1 = 16\,\text{MeV}/c^2,$$

$$\text{surface energy: } a_2 = 19,$$

$$\text{Coulomb energy: } a_3 = 0.72,$$

$$\text{symmetry energy: } a_4 = 28.$$

The symmetry energy favors $A = Z$. The term $\delta(A)$ is the "pairing energy" that gives small fluctuations.

Take $Z = N = A/2$, $m_n \approx m_p$, and neglect $\delta(A)$, so that

$$M = (m_p - a_1)A + a_2 A^{2/3} + \frac{a_3}{4} A^{5/3}.$$

Make a log plot of $M$ from $A = 1$ to $A = 10^4$, and indicate the range in which $M$ can be considered as extensive.

**1.9**   *The false vacuum:* In the theory of the "inflationary universe", a pin-point universe was created during the Big Bang in a "false" vacuum state characterized by a finite energy density $u_0$, while the true vacuum state should have zero energy density (Guth 1992).

An ordinary solid, too, has finite characteristic energy per unit volume, but it occupies a definite volume, given a total number of particles. If we put the solid in a box larger than that volume, it would not fill the box, but "rattle" in it. Not so for the false vacuum. It must have the same energy density whatever its volume, and this makes it a strange object indeed.

(a)   Imagine putting the false vacuum in a cylinder with piston, with a true vacuum outside. If we pull on the piston to increase the volume of the false vacuum by $dV$, show that the false vacuum performs work $dW = -u_0 \, dV$. Hence the false vacuum has negative pressure $P = -u_0$.

(b)   According to general relativity the radius $R$ of the universe obeys the equation

$$\frac{d^2 R}{dt^2} = -\frac{4\pi}{3c^2} G(u_0 - 3P)R,$$

where $c$ is the velocity of light, and

$$G = 6.673 \times 10^{-8} \text{ cm}^3/\text{g s}^2 = 6.707 \times 10^{-39} \text{ GeV}^{-2}\hbar c^3$$

is Newton's constant (gravitational constant). For the false vacuum, this reduces to

$$\frac{d^2 R}{dt^2} = \frac{R}{\tau^2},$$

$$\tau = \sqrt{\frac{3c^2}{8\pi G u_0}}.$$

Thus, the radius of the universe expands expands exponentially with time constant $\tau$. According to theory, $u_0 = (10^{15} \text{ GeV})^4/(\hbar c)^3 = 10^{96} \text{ erg/cm}^3$. Show that

$$\tau \approx 10^{-34} \text{ s}.$$

# Chapter 2

# Heat and entropy

## 2.1 The heat equations

Thermodynamics becomes a mathematical science when we regard the state functions, such as the internal energy, as continuous differentiable functions of the variables $P$, $V$ and $T$. The constraint imposed by the equation of state reduces the number of independent variables to two. We may consider the internal energy to be a function of any two of the variables. Under infinitesimal increments of the variables, we can write

$$
dU(P, V) = \left(\frac{\partial U}{\partial P}\right)_V dP + \left(\frac{\partial U}{\partial V}\right)_P dV,
$$

$$
dU(P, T) = \left(\frac{\partial U}{\partial P}\right)_T dP + \left(\frac{\partial U}{\partial T}\right)_P dT, \tag{2.1}
$$

$$
dU(V, T) = \left(\frac{\partial U}{\partial V}\right)_T dV + \left(\frac{\partial U}{\partial T}\right)_V dT,
$$

where a subscript on a partial derivative denotes the variable being held fixed. For example, $(\partial U/\partial P)_V$ is the derivative with respect to $P$ at constant $V$. These partial derivatives are thermodynamic coefficients to be taken from experiments.

The heat absorbed by the system can be obtained from the first law, written in the form $dQ = dU + P\,dV$:

$$
dQ = \left(\frac{\partial U}{\partial P}\right)_V dP + \left[\left(\frac{\partial U}{\partial V}\right)_P + P\right] dV,
$$

$$
dQ = \left(\frac{\partial U}{\partial P}\right)_T dP + \left(\frac{\partial U}{\partial T}\right)_P dT + P\,dV, \tag{2.2}
$$

$$
dQ = \left[\left(\frac{\partial U}{\partial V}\right)_T + P\right] dV + \left(\frac{\partial U}{\partial T}\right)_V dT.
$$

In the second of these equations, we must regard $V$ as a function of $P$ and $T$, and rewrite

$$dV = \left(\frac{\partial V}{\partial P}\right)_T dP + \left(\frac{\partial V}{\partial T}\right)_P dT. \tag{2.3}$$

The second equation then reads

$$dQ = \left[\left(\frac{\partial U}{\partial P}\right)_T + P\left(\frac{\partial V}{\partial P}\right)_T\right] dP + \left(\frac{\partial(U+PV)}{\partial T}\right)_P dT. \tag{2.4}$$

It is convenient to define a state function called the *enthalpy*:

$$H \equiv U + PV. \tag{2.5}$$

The heat equations in terms of $dQ$ are summarized below:

$$dQ = \left(\frac{\partial U}{\partial P}\right)_V dP + \left[\left(\frac{\partial U}{\partial V}\right)_P + P\right] dV,$$

$$dQ = \left[\left(\frac{\partial U}{\partial P}\right)_T + P\left(\frac{\partial V}{\partial P}\right)_T\right] dP + \left(\frac{\partial H}{\partial T}\right)_P dT, \tag{2.6}$$

$$dQ = \left[\left(\frac{\partial U}{\partial V}\right)_T + P\right] dV + \left(\frac{\partial U}{\partial T}\right)_V dT.$$

We can immediately read off the heat capacities at constant $V$ and $P$:

$$C_V = \left(\frac{\partial U}{\partial T}\right)_V, \qquad C_P = \left(\frac{\partial H}{\partial T}\right)_P. \tag{2.7}$$

These are useful because they express the heat capacities as derivatives of state functions.

## 2.2   Applications to ideal gas

We now use the first law to deduce some properties of the ideal gas. Joule performed a classic experiment on free expansion, as illustrated in Fig. 2.1. A thermally insulated ideal gas was allowed to expand freely into an insulated chamber, which was initially vacuous . After a new equilibrium was established, in which the gas fills both compartments, the final temperature was found to be the same as the initial temperature. We can draw the following conclusions:

- $\Delta W = 0$, since the gas pushes into a vacuum.
- $\Delta Q = 0$, since the temperature was unchanged.
- $\Delta U = 0$, by the first law.

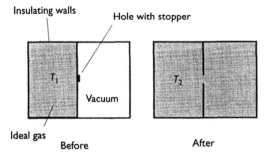

Figure 2.1 Free expansion of an ideal gas.

Choosing $V$ and $T$ as independent variables, we conclude that $U(V_1,T) = U(V_2,T)$, that is, $U$ is independent of $V$:

$$U = U(T). \tag{2.8}$$

Of course, $U$ is proportional to the total number of particles $N$, which has been kept constant.

The heat capacity at constant volume can now be written as a total derivative:

$$C_V = \left(\frac{\partial U}{\partial T}\right)_V = \frac{dU}{dT}. \tag{2.9}$$

Assuming that $C_V$ is a constant, we can integrate the above to obtain

$$U(T) = \int C_V \, dT = C_V T, \tag{2.10}$$

where the constant of integration has been set to zero by defining $U = 0$ at $T = 0$. It follows that

$$\begin{aligned} C_P &= \left(\frac{\partial H}{\partial T}\right)_P = \left(\frac{\partial (U + PV)}{\partial T}\right)_P \\ &= \frac{dU}{dT} + \frac{\partial (NkT)}{\partial T} \\ &= C_V + Nk. \end{aligned} \tag{2.11}$$

Thus, for an ideal gas,

$$C_P - C_V = Nk. \tag{2.12}$$

We now work out the equation governing a reversible adiabatic transformation. Setting $dQ = 0$, we have $dU = -P \, dV$. Since $dU = C_V \, dT$, we obtain

$$C_V \, dT + P \, dV = 0. \tag{2.13}$$

Using the equation of state $PV = NkT$, we can write

$$dT = \frac{d(PV)}{Nk} = \frac{P\,dV + V\,dP}{Nk}. \qquad (2.14)$$

Thus,

$$C_V(P\,dV + V\,dP) + NkP\,dV = 0,$$
$$C_V V\,dP + (C_V + Nk)P\,dV = 0, \qquad (2.15)$$
$$C_V V\,dP + C_P P\,dV = 0,$$

or

$$\frac{dP}{P} + \gamma\frac{dV}{V} = 0, \qquad (2.16)$$

where

$$\gamma \equiv \frac{C_P}{C_V}. \qquad (2.17)$$

Assuming that $\gamma$ is a constant, an integration yields

$$\ln P = -\gamma \ln V + \text{Constant}, \qquad (2.18)$$

or

$$PV^\gamma = \text{Constant}. \qquad (2.19)$$

Using the equation of state, we can rewrite this in the equivalent form

$$TV^{\gamma-1} = \text{Constant}. \qquad (2.20)$$

Since $\gamma > 1$ according to (2.12), an adiabatic path has a steeper slope than an isotherm in a $P$–$V$ diagram, as depicted in Fig. 2.2.

## 2.3  Carnot cycle

In a cyclic transformation, the final state is the same as the initial state and, therefore, $\Delta U = 0$, because $U$ is a state function. A reversible cyclic process can be represented by a closed loop in the $P$–$V$ diagram. The area of the loop is the total work done by the system in one cycle. Since $\Delta U = 0$, it is also equal to the heat absorbed:

$$\Delta W = \Delta Q = \oint P\,dV = \text{Area enclosed}. \qquad (2.21)$$

A cyclic process converts work into heat, and returns the system to its original state. It acts as a heat engine, for the process can be repeated indefinitely. If the cycle is reversible, it runs as a refrigerator in reverse.

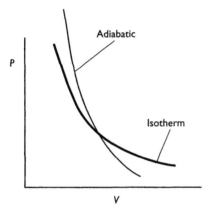

*Figure 2.2* Adiabatic line has a steeper slope than an isotherm.

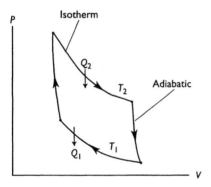

*Figure 2.3* Carnot cycle on P–V diagram of ideal gas.

A Carnot cycle is a reversible cycle bounded by two isotherms and two adiabatic lines. The working substance is arbitrary, but we illustrate it for an ideal gas in Fig. 2.3, where $T_2 > T_1$. The system absorbs heat $Q_2$ along the isotherm $T_2$, and rejects heat $Q_1$ along $T_1$, with $Q_1 > 0$ and $Q_2 > 0$. By the first law, the net work output is

$$W = Q_2 - Q_1. \tag{2.22}$$

In one cycle of operation, the system receives an amount of heat $Q_2$ from a hot reservoir, performs work $W$, and rejects "waste heat" $Q_1$ to a cold reservoir. The efficiency of the Carnot engine is defined as

$$\eta \equiv \frac{W}{Q_2} = 1 - \frac{Q_1}{Q_2}, \tag{2.23}$$

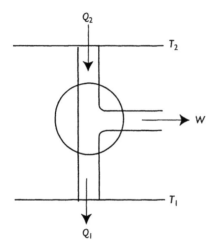

*Figure 2.4* Schematic representation of Carnot engine.

which is 100% if there is no waste heat, i.e. $Q_1 = 0$. But, as we shall see, the second law of thermodynamics states that this is impossible.

We represent the Carnot engine by the schematic diagram shown in Fig. 2.4, which emphasizes the fact that the working substance is irrelevant. By reversing the signs of $Q_1$ and $Q_2$, and thus $W$, we can run the engine in reverse as a Carnot refrigerator.

## 2.4  Second law of thermodynamics

The second law of thermodynamics expresses the common wisdom that "heat does not flow uphill". It is stated more precisely by Clausius:

> There does not exist a thermodynamic transformation whose sole effect is to deliver heat from a reservoir of lower temperature to a reservoir of higher temperature.

An equivalent statement is due to Kelvin:

> There does not exist a thermodynamic transformation whose sole effect is to extract heat from a reservoir and convert it entirely into work.

The important word is "sole". The processes referred to may be possible, but not without other effects. The logical equivalence of the two statements can be demonstrated by showing that the falsehood of one implies the falsehood of the other. Consider two heat reservoirs at respective temperatures $T_2$ and $T_1$, with $T_2 > T_1$.

(a) If the Kelvin statement were false, we could extract heat from $T_1$ and convert it entirely into work. We could then convert the work back to heat entirely, and deliver it to $T_2$ (there being no law against this). Thus, the Clausius statement would be negated.

(b) If the Clausius statement were false, we could let an amount of heat $Q_2$ flow uphill, from $T_1$ to $T_2$. We could then connect a Carnot engine between $T_2$ and $T_1$, to extract $Q_2$ from $T_2$, and return an amount $Q_1 < Q_2$ back to $T_1$. The net work output is $Q_2 - Q_1 > 0$. Thus, an amount of heat $Q_2 - Q_1$ is converted into work entirely, without any other effect. This would contradict the Kelvin statement. ∎

In the atomic view, heat transfer represents an exchange of energy residing in the random motion of the atoms. In contrast, the performance of work requires an organized action of the atoms. To convert heat entirely into work would mean that chaos spontaneously reverts to order. This is extremely improbable, for in the usual scheme of things, only one configuration corresponds to order, while all others lead to chaos. The second law is the thermodynamic way of expressing this idea.

## 2.5  Absolute temperature

The second law immediately implies that a Carnot engine cannot be 100% efficient, for otherwise all the heat absorbed from the upper reservoir would be converted into work in one cycle of operation. There is no other effect, since the system returns to its original state.

We can show that no engines working between two given temperatures can be more efficient than a Carnot engine. Since only two reservoirs are present, a Carnot engine simply means a reversible engine. What we assert then is that an irreversible engine cannot be more efficient than a reversible one.

Consider a Carnot engine $C$ and an engine $X$ (not necessarily reversible) working between the reservoirs $T_2$ and $T_1$, with $T_2 > T_1$, as shown in Fig. 2.5. We shall run $C$ in reverse, as a refrigerator $\bar{C}$, and feed the work output of $X$ to $\bar{C}$. Table 2.1 shows a balance sheet of heat transfer in one cycle of joint operation.

The total work output is

$$W_{\text{tot}} = (Q_2' - Q_1') - (Q_2 - Q_1). \tag{2.24}$$

Now arrange to have $Q_2' = Q_2$. Then, no net heat was extracted from the reservoir $T_2$, which can be ignored. An amount of heat $Q_1 - Q_1'$ was extracted from the reservoir $T_1$, and converted entirely into work with no other effect. This would violate the second law, unless $Q_1 \leq Q_1'$. Dividing both sides of this inequality by $Q_2$, and using the fact that $Q_2' = Q_2$, we have

$$\frac{Q_1}{Q_2} \leq \frac{Q_1'}{Q_2'}. \tag{2.25}$$

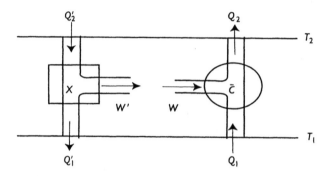

Figure 2.5 Driving a Carnot refrigerator $\bar{C}$ with an engine X.

Table 2.1  Balance sheet of
          heat transfer

| Engine | From $T_2$ | To $T_1$ |
|--------|-----------|----------|
| X | $Q_2'$ | $Q_1'$ |
| $\bar{C}$ | $-Q_2$ | $-Q_1$ |

Therefore, $1 - (Q_1/Q_2) \geq 1 - (Q_1'/Q_2')$, or

$$\eta_C \geq \eta_X. \tag{2.26}$$

As a corollary, all Carnot engines have the same efficiency, since X may be a Carnot engine. This shows that the Carnot engine is universal, in that it depends only on the temperatures involved, and not on the working substance.

We define the *absolute temperature* $\theta$ of a heat reservoir such that the ratio of the absolute temperatures of two reservoirs is given by

$$\frac{\theta_1}{\theta_2} \equiv \frac{Q_1}{Q_2} = 1 - \eta, \tag{2.27}$$

where $\eta$ is the efficiency of a Carnot engine operating between the two reservoirs. The advantage of this definition is that it is independent of the properties of any working substance. Since $Q_1 > 0$ according to the second law, the absolute temperature is bounded from below:

$$\theta > 0. \tag{2.28}$$

The *absolute zero*, $\theta = 0$, is a limiting value which we can never reach, again according to the second law.

The absolute temperature coincides with the ideal gas temperature defined by $T = PV/Nk$, as we can show by using an ideal gas as working substance in a Carnot engine. Thus, $\theta = T$, and we shall henceforth denote the absolute temperature by $T$.

The existence of absolute zero does not mean that the temperature scale terminates at the low end, for the scale is an open set without boundaries. It is a matter of convention that we call $T$ the temperature. We could have used $1/T$ as the temperature, and what is now absolute zero would be infinity instead.

## 2.6 Temperature as integrating factor

We can look upon the absolute temperature $T$ as the integrating factor that converts the inexact differential $dQ$ into an exact differential $dQ/T$. In order to show this, we first prove a theorem due to Clausius:

*In an arbitrary cyclic process P, the following inequality holds:*

$$\oint_P \frac{dQ}{T} \leq 0,\tag{2.29}$$

*where the equality holds if P is reversible.*

It should be emphasized that the cycle process need not be reversible. In order to prove the assertion, divide the cycle $P$ into $K$ segments labeled $i = 1, \ldots, K$. Let the $i$th segment be in contact with a reservoir of temperature $T_i$, from which it absorbs an amount of heat $Q_i$. The total work output of $P$ is, by the first law,

$$W = \sum_{i=1}^{K} Q_i.\tag{2.30}$$

Note that not all the $Q_i$ can be positive, for otherwise heat would have been converted into work with no other effect, in contradiction to the second law.

Imagine a reservoir at a temperature $T_0 > T_i$ (all $i$), with Carnot engines $C_i$ operating between $T_0$ and each of the temperatures $T_i$. The setup is illustrated schematically in Fig. 2.6. Suppose that, in one cycle of operation, the Carnot engine $C_i$ absorbs heat $Q_i^{(0)}$ from the $T_0$ reservoir, and rejects $Q_i$ to the $T_i$ reservoir. By definition of the absolute temperature, we have

$$\frac{Q_i^{(0)}}{Q_i} = \frac{T_0}{T_i}.\tag{2.31}$$

In one cycle of operation of the joint operations $\{P + C_1 + \cdots + C_K\}$,

- the reservoir $T_i$ experiences no net change, for it receives $Q_i$, and delivers the same to the system;
- the heat extracted from the $T_0$ reservoir is given by

$$Q_{tot} = \sum_{i=1}^{K} Q_i^{(0)} = T_0 \sum_{i=1}^{K} \frac{Q_i}{T_i};\tag{2.32}$$

- the total work output is

$$W_{\text{tot}} = W + \sum_{i=1}^{K} \left[ Q_i^{(0)} - Q_i \right] = \sum_{i=1}^{K} Q_i^{(0)} = Q_{\text{tot}}. \tag{2.33}$$

An amount of heat $Q_{\text{tot}}$ would be entirely converted into work with no other effect, and thus violate the second law, unless $Q_{\text{tot}} \leq 0$, or

$$\sum_{i=1}^{K} \frac{Q_i}{T_i} \leq 0. \tag{2.34}$$

In the limit $K \to \infty$, this becomes

$$\oint_P \frac{dQ}{T} \leq 0. \tag{2.35}$$

which proves the first part of the theorem.

If $P$ is reversible, we can run the operation $\{P + C_1 + \cdots + C_N\}$ in reverse. The signs for $Q_i$ are then reversed, and we conclude that $\oint_P dQ/T \geq 0$. Combining

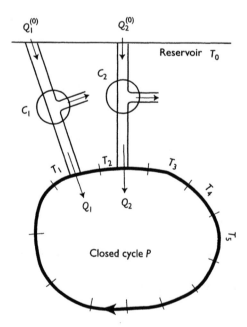

Figure 2.6 Construction to prove Clausius' theorem.

this with the earlier relation, which still holds, we have

$$\oint_P \frac{dQ}{T} = 0 \quad \text{(if } P \text{ is reversible).} \tag{2.36}$$

This completes the proof. ∎

A corollary to the theorem is that, for a reversible open path $P$, the integral $\int_P dQ/T$ depends only on the endpoints, and not on the particular path. In order to prove this, join the endpoints by a reversible path $P'$, whose reversal is denoted by $-P'$. The combined processes $P - P'$ then represents a closed reversible cycle. Therefore, $\int_{P-P'} dQ/T = 0$, or

$$\int_P \frac{dQ}{T} = \int_{P'} \frac{dQ}{T}. \tag{2.37}$$

This shows that $dQ$ divided by the absolute temperature is an exact differential. ∎

## 2.7 Entropy

The exact differential

$$dS = \frac{dQ}{T} \tag{2.38}$$

defines a state function $S$ called the entropy, up to an additive constant. The entropy difference between any two states $B$ and $A$ is given by

$$S(B) - S(A) \equiv \int_A^B \frac{dQ}{T} \quad \text{(along reversible path),} \tag{2.39}$$

where the integral extends along any *reversible* path connecting $B$ to $A$. The result is, of course, independent of the path, as long as such a reversible path exists.

What if we integrate along an irreversible path? Let $P$ be an arbitrary path from $A$ to $B$, reversible or not. Let $R$ be a reversible path with the same endpoints. The combined process $P - R$ is a closed cycle and, therefore, by Clausius' theorem, $\int_{P-R} dQ/T \leq 0$, or

$$\int_P \frac{dQ}{T} \leq \int_R \frac{dQ}{T}. \tag{2.40}$$

Since the right-hand side is the definition of the entropy difference between the final state $B$ and the initial state $A$, we have

$$S(B) - S(A) \geq \int_A^B \frac{dQ}{T}, \tag{2.41}$$

where the equality holds if the process is reversible. For an isolated system, which does not exchange heat with the external world, we have $dQ = 0$ and, therefore,

$$\Delta S \geq 0. \tag{2.42}$$

This means that the entropy of an isolated system never decreases, and it remains constant during a reversible transformation.

We emphasize the following points:

- The principle that the entropy never decreases applies to the "universe" consisting of a system and its environments. It does not apply to a non-isolated system, whose entropy may increase or decrease.
- Since the entropy is a state function, the entropy change of the system in going from state $A$ to state $B$ is $S_B - S_A$, regardless of the path, which may be reversible or irreversible. For an irreversible path, the entropy of the environment changes, whereas for a reversible path it does not.
- The entropy difference $S_B - S_A$ is not necessarily equal to the integral $\int_A^B dQ/T$. It is equal to the integral only if the path from $A$ to $B$ is reversible. Otherwise, it is generally *larger* than the integral.

## 2.8  Entropy of ideal gas

We can calculate the entropy of an ideal gas as a function of $V$ and $T$ by integrating $dS = dQ/T$. In Fig. 2.7, we approach point $A$ along two alternative paths, with $V$ kept fixed along path 1, and $T$ kept fixed along path 2. Along path 1, we have $\int dQ/T = C_V \int dT/T$ and, hence,

$$S(V,T) = S(V,T_0) + C_V \int_{T_0}^{T} \frac{dT}{T} = S(V,T_0) + C_V \ln \frac{T}{T_0}. \tag{2.43}$$

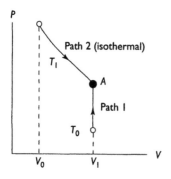

Figure 2.7 Calculating the entropy at point A.

In order to determine $S(V, T_0)$, we integrate $dS = dQ/T$ along the isothermal path 2. Since $dU = 0$, we have

$$dQ = dW = P\,dV = NkT\,\frac{dV}{V}. \tag{2.44}$$

Thus,

$$S(V, T) = S(V_0, T) + Nk \int_{V_0}^{V} \frac{dV}{V} = S(V_0, T) + Nk \ln \frac{V}{V_0}. \tag{2.45}$$

Comparing the two expressions for $S(V, T)$, we conclude that

$$S(V, T_0) = C_0 + Nk \ln V, \tag{2.46}$$

where $C_0$ is an arbitrary constant. Absorbing the constant $V_0$ into $C_0$, we can write

$$S(V, T) = C_0 + Nk \ln V + C_V \ln T. \tag{2.47}$$

For a monatomic gas, we have $C_V = \frac{3}{2}Nk$ and, hence,

$$S(V, T) = C_0 + Nk \ln (VT^{3/2}). \tag{2.48}$$

The constant $C_0$ is to be determined, and it could depend on $N$. A problem arises when $N$ changes, for $S$ behaves like $N \ln V$, and is not extensive, unless somehow $C_0$ contains a term $-Nk \ln N$. As we shall see in Section 12.6, quantum mechanics supplies the remedy by fixing $C_0$. The corrected entropy is given by the *Sacker–Tetrode equation*

$$S = Nk \left[ \frac{5}{2} - \ln \left( n\lambda^3 \right) \right], \tag{2.49}$$

where $n = N/V$ and $\lambda = \sqrt{2\pi\hbar^2/mkT}$. The latter is called the *thermal wavelength*, which is of order of the deBroglie wavelength of a particle with energy $kT$.

## 2.9 The limits of thermodynamics

Thermodynamics is an elegant theory and a very useful practical tool. The self-consistency of the mathematical structure has been demonstrated through axiomatic formulation. However, confrontation with experiments indicates that thermodynamics is valid only insofar as the atomic structure of matter can be ignored.

In the atomic picture, thermodynamic quantities are subject to small fluctuations. The second law of thermodynamics is true only on the macroscopic scale, when such fluctuations can be neglected. It is constantly being violated on the atomic scale.

Taking the second law as absolute would lead to the conclusion that the entropy of the universe must forever increase, leading towards an ultimate "heat death". Needless to say, this ceased to be imperative in the atomic picture. We shall discuss the meaning of macroscopic irreversibility in more detail in Section 6.9.

## Problems

**2.1**   An ideal gas undergoes a reversible transformation along the path $P = aV^b$, where $a$ and $b$ are constants, with $a > 0$. Find the heat capacity $C$ along this path.

**2.2**   The temperature in a lake is 300 K at the surface, and 290 K at the bottom. What is the maximum energy that can be extracted thermally from 1 g of water by exploiting the temperature difference?

**2.3**   A nuclear power plant generates 1 MW of power at a reactor temperature of 600°F. It rejects waste heat into a nearby river, with a flow rate of 6,000 ft³/s, and an upstream temperature of 70°F. The power plant operates at 60% of maximum possible efficiency. Find the temperature rise of the river.

**2.4**   A slow-moving stream carrying hot spring water at temperature $T_2$ joins a sluggish stream of glacial water at temperature $T_1$. The water downstream has a temperature $T$ between $T_2$ and $T_1$, and flows much faster, at a velocity $v$. Assume that the specific heat of water is a constant, and that the entropy of water has the same temperature dependence as that of an ideal gas.

(a) Find $v$, neglecting the velocities of the input streams. (*Hint*: The net change in thermal energy was converted to kinetic energy.)

(b) Find the lower bound on $T$ and the upper bound on $v$ imposed by the second law of thermodynamics. (*Hint*: The total entropy cannot decrease.)

**2.5**   A cylinder of cross-sectional area $A$ is divided into two chambers 1 and 2, by means of a frictionless piston. The piston as well as the walls of the chambers are heat-insulating, and the chambers initially have equal length $L$. Both chambers are filled with 1 mol of helium gas, with initial pressures $2P_0$ and $P_0$, respectively. The piston is then allowed to slide freely, whereupon the gas in chamber 1 pushes the piston a distance $a$ to equalize the pressures to $P$.

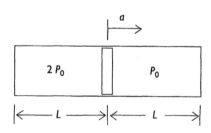

(a) Find the distance $a$ traveled by the piston.

(b) If $W$ is the work done by the gas in chamber 1, what are the final temperatures $T_1$ and $T_2$ in the two chambers? What is the final pressure $P$?

(c) Find $W$, the work done.

**2.6**  The equation of state of radiation is $PV = U/3$. Stefan's law gives $U/V = \sigma T^4$, with $\sigma = \pi^2 k^4/(15\hbar^3 c^3)$.

(a)  Find the entropy of radiation as a function of $V$ and $T$.

(b)  During the Big Bang, the radiation, initially confined within a small region, expands adiabatically and cools down. Find the relation between the temperature $T$ and the radius of the universe $R$.

**2.7**  Put an ideal gas through a Carnot cycle and show that the efficiency is $\eta = 1 - T_1/T_2$, where $T_2$ and $T_1$ are the ideal-gas temperatures of the heat reservoirs. This shows that the ideal-gas temperature coincides with the absolute temperature.

**2.8**  The Diesel cycle is illustrated in the accompanying diagram. Let $r = V_1/V_2$ (compression ratio) and $r_c = V_3/V_2$ (cutoff ratio). Assuming that the working substance is an ideal gas with $C_P/C_V = \gamma$, find the efficiency of the cycle.

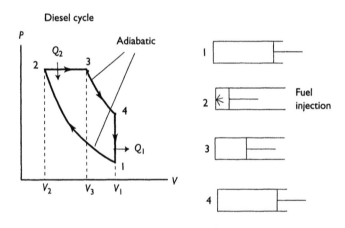

Diesel cycle

**2.9**  The Otto cycle is shown in the accompanying diagram, where $r = V_1/V_2$ is the compression ratio. The working substance is an ideal gas. Find the efficiency.

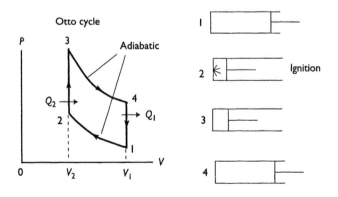

Otto cycle

**2.10**   An ideal gas undergoes a cyclic transformation *abca* (see sketch) such that

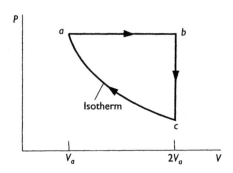

- *ab* is at constant pressure, with $V_b = 2V_a$,
- *bc* is at constant volume,
- *ca* is at constant temperature.

Find the efficiency of this cycle, and compare it with that of the Carnot cycle operating between the highest and lowest available temperatures.

**2.11**   A monatomic classical ideal gas is taken from point *A* to point *B* in the *P–V* diagram shown in the accompanying sketch, along three different reversible paths *ACB*, *ADB* and *AB*, with $P_2 = 2P_1$ and $V_2 = 2V_1$. The thick lines are isotherms:

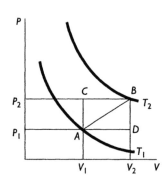

(a) Find the heat supplied to the gas in each of the three transformations, in terms of $Nk$, $P_1$ and $T_1$.

(b) What is the heat capacity along the path *AB*?

(c) Find the efficiency of the heat engine based on the closed cycle *ACBD*.

**2.12**   Here is a device that allegedly violates the second law of thermodynamics. Consider the surface of rotation shown in heavy lines in the accompanying sketch. It is made up of parts of the surfaces of two confocal ellipsoids of revolution, and that of a sphere. The inside surface is a perfect mirror. The foci are labeled *A* and *B*.

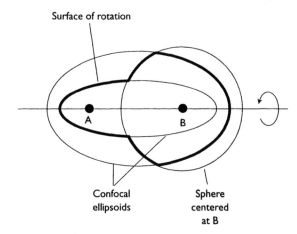

Surface of rotation

A

B

Confocal
ellipsoids

Sphere
centered
at B

The argument goes as follows: If two black bodies of equal temperature are initially placed at $A$ and $B$, then all the radiation from $A$ will reach $B$, but not vice versa, because radiation from $B$ hitting the spherical surface will be reflected back. Therefore, the temperature of $B$ will increase spontaneously, while that of $A$ will decrease spontaneously, and this would violate the second law.

(a)  Why would a spontaneous divergence of temperatures violate the second law?

(b)  Is the assertion true for physical black bodies?

# Chapter 3

# Using thermodynamics

## 3.1  The energy equation

In Chapter 2, we obtained the thermodynamic equations in terms of $dQ$ which give the amount of heat absorbed by a system when the independent variables change. However, the formulas involve derivatives of the internal energy, which is not directly measurable. We can obtain more practical results by rewriting the equations by exploiting the fact that $dS = dQ/T$ is an exact differential. We illustrate the method using (2.6), with $T$ and $V$ as independent variables:

$$dQ = T\,dS = C_V\,dT + \left[\left(\frac{\partial U}{\partial V}\right)_T + P\right]dV. \tag{3.1}$$

Dividing both sides by $T$, we obtain an exact differential:

$$dS = \frac{C_V}{T}\,dT + \frac{1}{T}\left[\left(\frac{\partial U}{\partial V}\right)_T + P\right]dV, \tag{3.2}$$

which must be of the form

$$dS = \frac{\partial S}{\partial T}\,dT + \frac{\partial S}{\partial V}\,dV. \tag{3.3}$$

Here we have suppressed the subscripts on partial derivatives. Thus, we can identify

$$\frac{\partial S}{\partial T} = \frac{C_V}{T},$$

$$\frac{\partial S}{\partial V} = \frac{1}{T}\left[\frac{\partial U}{\partial V} + P\right]. \tag{3.4}$$

Since differentiation is a commutative operation, we have

$$\frac{\partial}{\partial V}\frac{\partial S}{\partial T} = \frac{\partial}{\partial T}\frac{\partial S}{\partial V}. \tag{3.5}$$

Hence

$$\frac{\partial}{\partial V}\frac{C_V}{T} = \frac{\partial}{\partial T}\left[\frac{1}{T}\left(\frac{\partial U}{\partial V} + P\right)\right]. \tag{3.6}$$

The left-hand side can be written as $T^{-1}\partial C_V/\partial V$, since $T$ is kept fixed for the differentiation. Using $C_V = \partial U/\partial T$, we can rewrite the last equation in the form

$$\frac{1}{T}\frac{\partial}{\partial V}\frac{\partial U}{\partial T} = -\frac{1}{T^2}\left[\frac{\partial U}{\partial V} + P\right] + \frac{1}{T}\left[\frac{\partial}{\partial V}\frac{\partial U}{\partial T} + \frac{\partial P}{\partial T}\right]. \tag{3.7}$$

After canceling identical terms on both sides, we obtain the *energy equation*

$$\left(\frac{\partial U}{\partial V}\right)_T = T\left(\frac{\partial P}{\partial T}\right)_V - P, \tag{3.8}$$

where we have restored the subscripts. The derivative of the internal energy is now expressed in terms of readily measurable quantities.

In Chapter 2, from Joule's free expansion experiment, we deduced that the internal energy for an ideal gas depends only on the temperature, and not the volume, i.e. $(\partial U/\partial V)_T = 0$. Now we can show that this is implied by the second law, through the energy equation. Using the equation of state for the ideal gas, we have

$$\left(\frac{\partial P}{\partial T}\right)_V = \frac{P}{T} \text{ (ideal gas).} \tag{3.9}$$

Therefore, by the energy equation,

$$\left(\frac{\partial U}{\partial V}\right)_T = 0 \text{ (ideal gas).} \tag{3.10}$$

## 3.2  Some measurable coefficients

The energy equation relates the experimentally inaccessible quantity $(\partial U/\partial V)_T$ to $(\partial P/\partial T)_V$. The latter, in turn, can be related to other thermodynamic coefficients. Using the chain rule for partial derivatives (see Appendix), we can write

$$\left(\frac{\partial P}{\partial T}\right)_V = -\frac{1}{(\partial T/\partial V)_P\,(\partial V/\partial P)_T} = \frac{(\partial V/\partial T)_P}{-(\partial V/P)_T} = \frac{\alpha}{\kappa_T}, \tag{3.11}$$

where $\alpha$ and $\kappa_T$ are among some directly measurable coefficients:

$$\alpha = \frac{1}{V}\left(\frac{\partial V}{\partial T}\right)_P \text{ (coefficient of thermal expansion),}$$

$$\kappa_T = -\frac{1}{V}\left(\frac{\partial V}{\partial P}\right)_T \text{ (isothermal compressibility),} \tag{3.12}$$

$$\kappa_S = -\frac{1}{V}\left(\frac{\partial V}{\partial P}\right)_S \text{ (adiabatic compressibility).}$$

Substituting the new form of $(\partial P/\partial T)_V$ into $(\partial U/\partial V)_T$, and then the latter into equation (3.1), we obtain

$$T \, dS = C_V \, dT + \frac{\alpha T}{\kappa_T} \, dV. \tag{3.13}$$

This "$T \, dS$ equation" gives the heat absorbed in terms of directly measurable coefficients. Using $T$ and $P$ as independent variables, we have

$$T \, dS = C_P \, dT - \alpha TV \, dP. \tag{3.14}$$

If $V$ and $P$ are used as independent variables, we rewrite $dT$ in terms of $dV$ and $dP$:

$$dT = \left(\frac{\partial T}{\partial V}\right)_P dV + \left(\frac{\partial T}{\partial P}\right)_V dP = \frac{1}{\alpha V} \, dV + \frac{\kappa_T}{\alpha} \, dP. \tag{3.15}$$

In summary, the $T \, dS$ equations are

$$\begin{aligned}
T \, dS &= C_V \, dT + \frac{\alpha T}{\kappa_T} \, dV, \\
T \, dS &= C_P \, dT - \alpha TV \, dP, \\
T \, dS &= \frac{C_P}{\alpha V} \, dV + \left(\frac{C_P \kappa_T}{\alpha} - \alpha TV\right) dP.
\end{aligned} \tag{3.16}$$

## 3.3   Entropy and loss

In Chapter 2, we showed that the entropy of an isolated system remains constant during a reversible process, but it increases during an irreversible process, when the system goes out of thermal equilibrium. Useful energy is lost in the process, and the increase of entropy measures the loss.

In order to illustrate this point, consider 1 mol of an ideal gas at temperature $T$, expanding from volume $V_1$ to $V_2$. Let us compare the entropy change for a reversible isothermal expansion and an irreversible free expansion. The processes are schematically depicted in Fig. 3.1. They have the same initial and final states, as indicated on the P–V diagram in Fig. 3.2. However, the path for the free expansion cannot be represented on the diagram, because the pressure is not well defined.

In the reversible isothermal expansion, $\Delta U = 0$ because $U$ depends only on the temperature for an ideal gas. By the first law, we have

$$\Delta Q = \Delta W = \int_{V_1}^{V_2} \frac{RT}{V} \, dV = RT \ln \frac{V_2}{V_1}. \tag{3.17}$$

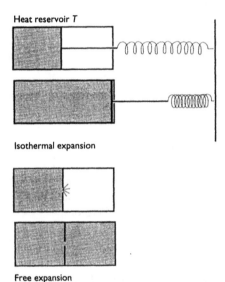

Heat reservoir $T$

Isothermal expansion

Free expansion

Figure 3.1 Reversible isothermal expansion and irreversible free expansion. In the former case, the temperature is maintained by a heat reservoir, and the work done is stored externally.

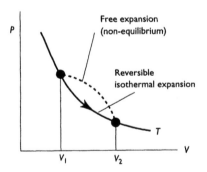

Free expansion
(non-equilibrium)

Reversible
isothermal expansion

Figure 3.2 Isothermal expansion and free expansion. The latter cannot be represented by a path, because pressure is not well defined during the process.

The work done $\Delta W$ is stored in a spring attached to the moving wall, and it can be used to compress the gas back to the initial state. The entropy change of the gas is given by

$$(\Delta S)_{\text{gas}} = \int \frac{dQ}{T} = \frac{\Delta Q}{T} = R \ln \frac{V_2}{V_1}. \tag{3.18}$$

Since the heat reservoir delivers an amount of heat $\Delta Q$ to the gas at temperature $T$, its entropy change is

$$(\Delta S)_{\text{reservoir}} = -\Delta Q/T = -R \ln \frac{V_2}{V_1}. \tag{3.19}$$

Thus, there is no change in the entropy of the "universe" made up of the system and its environment:

$$(\Delta S)_{\text{universe}} \equiv (\Delta S)_{\text{gas}} + (\Delta S)_{\text{reservoir}} = 0. \tag{3.20}$$

In the free expansion, $\Delta W = 0$ because the gas expands into a vacuum. Since the temperature of the gas does not change, according to Joule's experiment, there is no heat transfer between the gas and its environment. Therefore,

$$(\Delta S)_{\text{reservoir}} = 0. \tag{3.21}$$

The entropy change of the gas is the same as that calculated before, since it depends only on the initial and final states:

$$(\Delta S)_{\text{gas}} = R \ln \frac{V_2}{V_1}. \tag{3.22}$$

Therefore,

$$(\Delta S)_{\text{universe}} = R \ln \frac{V_2}{V_1}. \tag{3.23}$$

Had the transformation proceeded reversibly, the gas could have performed work amounting to

$$\Delta W = RT \ln \frac{V_2}{V_1} = T(\Delta S)_{\text{universe}}. \tag{3.24}$$

The increase in total entropy is a reflection of the loss of useful energy.

In heat conduction, an amount of heat $\Delta Q$ is directly transferred from a hotter reservoir $T_2$ to a cooler one $T_1$, and the entropy change of the universe is

$$(\Delta S)_{\text{universe}} = \frac{\Delta Q}{T_2} - \frac{\Delta Q}{T_1} > 0. \tag{3.25}$$

This shows that heat conduction is always irreversible. The only reversible way to transfer heat from $T_2$ to $T_1$ is to connect a Carnot engine between the two reservoirs, so that the work output can be used to reverse the process.

## 3.4  The temperature–entropy diagram

We can use the entropy $S$ as a thermodynamic variable and represent a thermodynamic process on a $T$–$S$ diagram, instead of the $P$–$V$ diagram. In such a representation, adiabatic lines are vertical lines, and the area under a path is the heat absorbed: $\int T \, dS = \Delta Q$.

A Carnot cycle is a rectangle, as illustrated in Fig. 3.3. The heat absorbed is equal to the area $A + B$. The heat rejected is equal to the area $B$, and the total work output is $A$. The efficiency is, therefore,

$$\eta = \frac{A}{A + B}. \tag{3.26}$$

The $T$–$S$ diagram is helpful in analyzing non-Carnot cycles, as illustrated in Fig. 3.4. The cycle in the left panel is equivalent to two Carnot cycles labeled 1 and 2 in the right panel, running simultaneously. The efficiencies of these cycles

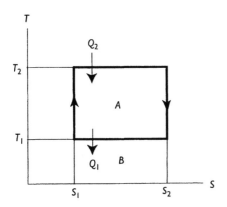

*Figure 3.3* Carnot cycle on a $T$–$S$ diagram

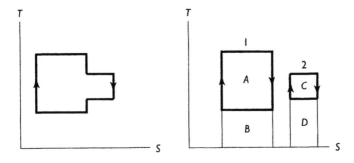

*Figure 3.4* The non-Carnot cycle on the left panel is equivalent to the two cycles shown on the right panel.

are given by the ratios of the areas indicated in the sketch:

$$\eta_1 = \frac{A}{A+B}, \qquad \eta_2 = \frac{C}{C+D}. \tag{3.27}$$

The overall cycle takes in as heat input $A + B + C + D$ and outputs total work $A + C$. Thus, the overall efficiency is

$$\eta = \frac{A+C}{A+B+C+D}. \tag{3.28}$$

It is clear that $\eta < \eta_1$ and $\eta < \eta_2$. Thus,

$$\eta < \min(\eta_1, \eta_2). \tag{3.29}$$

This shows that the largest and smallest available temperatures must be used to obtain the highest efficiency possible.

## 3.5  Condition for equilibrium

The first law states $\Delta U = \Delta Q - P \Delta V$. Using Clausius' theorem, $\Delta Q \leq T \Delta S$, we have

$$\Delta U \leq T \Delta S - \Delta W. \tag{3.30}$$

Thus, $\Delta U \leq 0$ for a system with $\Delta S = \Delta W = 0$. This means that the internal energy will seek the lowest possible value, when the system is thermally and mechanically isolated. For infinitesimal reversible changes, we have

$$dU = T \, dS - P \, dV. \tag{3.31}$$

Thus, the natural variables for $U$ are $S$ and $V$. If the function $U(S, V)$ is known, we can obtain all thermodynamic properties through the formulas

$$P = -\left(\frac{\partial U}{\partial V}\right)_S, \qquad T = \left(\frac{\partial U}{\partial S}\right)_V. \tag{3.32}$$

These are known as Maxwell relations.

## 3.6  Helmholtz free energy

It is difficult to manipulate $S$ and $V$ in the laboratory, but it is far easier to change $T$ and $V$. It is thus natural to ask, "What is the equilibrium condition at constant $T$, $V$?"

In order to answer this question, we go back to the inequality $\Delta U \leq T\Delta S - \Delta W$. If $T$ is kept constant, we can rewrite it in the form

$$\Delta W \leq -\Delta(U - TS). \tag{3.33}$$

If $\Delta W = 0$, then $(U - TS) \leq 0$. This motivates us to define a new thermodynamic function, the *Helmholtz free energy* (or simply *free energy*):

$$A \equiv U - TS. \tag{3.34}$$

The earlier inequality now reads

$$\Delta A \leq -\Delta W. \tag{3.35}$$

If $\Delta W = 0$, then $\Delta A \leq 0$. The equilibrium condition for a mechanically isolated body at constant temperature is that the free energy be minimum.

For infinitesimal reversible transformations, we have $dA = dU - T\,dS - S\,dT$. Using the first law, we can reduce this to

$$dA = -P\,dV - S\,dT. \tag{3.36}$$

If we know the function $A(T, V)$, then all thermodynamic properties can be obtained through the Maxwell relations

$$P = -\left(\frac{\partial A}{\partial V}\right)_T, \qquad S = -\left(\frac{\partial A}{\partial T}\right)_V. \tag{3.37}$$

The first of these reduces to the intuitive relation $P = -\partial U / \partial V$ at absolute zero.

In order to illustrate the minimization of free energy, consider the arrangement shown in Fig. 3.5. An ideal gas is contained in a cylinder divided into two compartments of volumes $V_1$ and $V_2$, respectively, with a dividing wall that can slide without friction. The entire system is in thermal contact with a heat reservoir of temperature $T$. As the partition slides, the total volume $V = V_1 + V_2$ as well as the temperature $T$ remain constant. Intuitively, we know that the partition will slide to such a position as to equalize the pressures on both sides. How can we show this purely on thermodynamic grounds? The answer is that we must minimize the

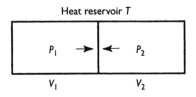

Figure 3.5 The sliding partition will come to rest at such a position as to minimize the free energy. This implies equalization of pressure: $P_1 = P_2$.

free energy. This means that, when equilibrium is established, any small displace of the partition will produce no change in the free energy to first order, i.e.

$$\delta A = \frac{\partial A}{\partial V_1}\,\delta V_1 + \frac{\partial A}{\partial V_2}\,\delta V_2 = 0 \qquad (3.38)$$

with the constraint $\delta V_1 + \delta V_2 = 0$. Thus, the condition for equilibrium is

$$\left(\frac{\partial A}{\partial V_1} - \frac{\partial A}{\partial V_2}\right)\delta V_1 = 0. \qquad (3.39)$$

Since $T$ is constant, the partial derivatives give the pressures; hence, $P_1 = P_2$.

## 3.7   Gibbs potential

We have seen that the thermodynamic properties of a system can be obtained from the function $U(S, V)$ or from $A(V, T)$, depending on the choice of independent variables. The replacement of $U$ by $A = U - TS$ was motivated by the fact that $dU = T\,dS - P\,dV$, and we want to replace the term $T\,dS$ by $S\,dT$. This is an example of a *Legendre transformation*.

Let us now consider $P$ and $T$ as independent variables. We introduce the Gibbs potential $G$, by making a Legendre transformation on $A$,

$$G \equiv A + PV. \qquad (3.40)$$

Then, $dG = dA + P\,dV + V\,dP = -S\,dT - P\,dV + P\,dV + V\,dP$, or

$$dG = -S\,dT + V\,dP. \qquad (3.41)$$

The condition for equilibrium at constant $T$, and $P$ is that $G$ be at a minimum. We now have further Maxwell relations

$$V = \left(\frac{\partial G}{\partial P}\right)_T, \qquad S = -\left(\frac{\partial G}{\partial T}\right)_P. \qquad (3.42)$$

The Gibbs potential is useful in describing chemical processes, which usually take place under constant atmospheric pressure.

## 3.8   Maxwell relations

The following basic functions are related to one another through Legendre transformations:

$$\begin{aligned}
& U(S, V)\text{:}\ dU = T\,dS - P\,dV, \\
& A(V, T)\text{:}\ dA = -S\,dT - P\,dV, \\
& G(P, T)\text{:}\ dG = -S\,dT + V\,dP, \\
& H(S, P)\text{:}\ dH = T\,dS + V\,dP.
\end{aligned} \qquad (3.43)$$

Each function is expressed in terms of its natural variables. When these variables are held fixed, the corresponding function is at a minimum in thermal equilibrium. Thermodynamic functions can be obtained through the Maxwell relations summarized in the diagram in Fig. 3.6.

## 3.9 Chemical potential

So far, in thermodynamic transformations, we have kept the number of particles $N$ constant. When $N$ does change, the first law is generalized to the form

$$dU = dQ - dW + \mu\, dN, \tag{3.44}$$

where $\mu$ is called the *chemical potential*, the energy needed to add one particle to a thermally and mechanically isolated system. For a gas–liquid system, we have

$$dU = T\, dS - P\, dV + \mu\, dN. \tag{3.45}$$

The change in free energy is given by

$$dA = -S\, dT - P\, dV + \mu\, dN, \tag{3.46}$$

which gives the Maxwell relation

$$\mu = \left(\frac{\partial A}{\partial N}\right)_{V,T}. \tag{3.47}$$

Similarly, for processes at constant $P$ and $T$, we consider the change in the Gibbs potential:

$$dG = -S\, dT - V\, dP + \mu\, dN \tag{3.48}$$

Figure 3.6 Mnemonic diagram summarizing the Maxwell relations. Each quantity at the center of a row or column is flanked by its natural variables. The partial derivative with respect to one variable, with the other kept fixed, is arrived at by following the diagonal line originating from that variable. Attach a minus sign if you go against the arrow.

and obtain

$$\mu = \left(\frac{\partial G}{\partial N}\right)_{P,T}. \tag{3.49}$$

## Problems

**3.1**   We derive some useful thermodynamic relations in this problem.

(a)   The $T\,dS$ equations remain valid when $dS = 0$. Exploiting this fact, express $C_V$ and $C_P$ in terms of adiabatic derivatives and show that

$$\frac{C_P}{C_V} = \frac{\kappa_T}{\kappa_S}.$$

(b)   Equate the right-hand side of the two $T\,dS$ equations, and then use $P$ and $V$ as independent variables. From this, derive the relation

$$C_P - C_V = \frac{\alpha^2 TV}{\kappa_T}.$$

(c)   Using the Maxwell relations, show that

$$C_V = -T\left(\frac{\partial^2 A}{\partial T^2}\right)_V, \qquad C_P = -T\left(\frac{\partial^2 G}{\partial T^2}\right)_P.$$

**3.2**   When the number of particles changes in a thermodynamic transformation, it is important to use the correct form of entropy for an ideal gas, as given by the Sacker–Tetrode equation (2.49).

(a)   Use the Sacker–Tetrode equation to calculate $A(V,T)$ and $G(P,T)$ for an ideal gas. Show, in particular, that

$$A(V,T) = NkT[\ln(n\lambda^3) - 1],$$

where $n$ is the density, and $\lambda = \sqrt{2\pi\hbar^2/mkT}$ is the thermal wavelength.

(b)   Obtain the chemical potential for an ideal gas from $(\partial A/\partial N)_{V,T}$ and $(\partial G/\partial N)_{P,T}$. Show that you get the same answer

$$\mu = kT\ln(n\lambda^3).$$

**3.3**   A glass flask with a long narrow neck, of small cross-sectional area $A$, is filled with 1 mol of a dilute gas with $C_P/C_V = \gamma$, at temperature $T$. A glass bead of mass $m$ fits snugly into the neck, and can slide along the neck without friction.

Find the frequency of small oscillations of the bead about its equilibrium position. This gives a method to measure $\gamma$.

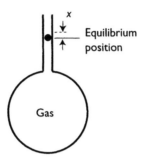

**3.4** A cylinder with insulating walls is divided into two equal compartments by means of an insulating piston of mass $M$, which can slide without friction. The cylinder has cross section $A$, and the compartments are of length $L$. Each compartment contains 1 mol of a classical ideal gas with $C_P/C_V = \gamma$, at temperature $T_0$ (see sketch).

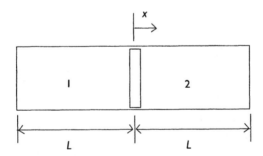

(a) Suppose the piston is adiabatically displaced a small distance $x \ll L$. Calculate, to first order in $x$, the pressures $P_1, P_2$ and temperatures $T_1, T_2$ in the two chambers.

(b) Find the frequency of small adiabatic oscillations of the piston about its equilibrium position.

(c) Now suppose that the piston has a small heat conductivity, so that heat flows from 1 to 2 at the rate $dQ/dt = K\Delta T$, where $K$ is very small, and $\Delta T = T_1 - T_2$. Find the rate of increase of the entropy of the universe.

(d) Entropy generation implies energy dissipation, which damps the oscillation. Calculate the energy dissipated per cycle.

**3.5**   A liquid has an equilibrium density corresponding to specific volume $v_0(T)$. Its free energy can be represented by

$$A(V, T) = Na_0(T)[v_0(T) - v]^2 - Nf(T),$$

where $v = V/N$.

(a) Find the equation of state $P(v, T)$ of the liquid.
(b) Calculate the isothermal compressibility $\kappa_T$ and the coefficient of thermal expansion, $\alpha$.
(c) Find the chemical potential.

*Note*: For $v > v_0$ the pressure becomes negative and, therefore, unphysical. See Problem 4.4 for remedy.

**3.6**   A mixture of two ideal gases undergoes an adiabatic transformation. The gases are labeled 1 and 2 Their densities and specific heats are denoted by $n_j$, $C_{Vj}$, $C_{Pj}$ ($j = 1, 2$). Show that the pressure $P$ and volume $V$ of the system obeys the constraint $PV^\xi = $ constant, where

$$\xi = \frac{n_1 C_{P1} + n_2 C_{P2}}{n_1 C_{V1} + n_2 C_{V2}}.$$

(*Hint*: The entropy of the system does not change, but those of the components do. For the entropy change of an ideal gas, use $\Delta S = Nk \ln (V_f/V_i) + NC_V \ln (T_f/T_i)$, where $f$ and $i$ denote final and initial values.)

**3.7**   Two thin disks of metal were at temperatures $T_2$ and $T_1$, respectively, with $T_2 > T_1$. They are brought into thermal contact on their flat surfaces, and came to equilibrium under atmospheric pressure, in thermal isolation,

(a) Find the final temperature.
(b) Find the increase in entropy of the universe.

**3.8**   The thermodynamic variables for a magnetic system are $H, M$ and $T$, where $H$ is the magnetic field, $M$ the magnetization and $T$ the absolute temperature. The magnetic work is $dW = -H \, dM$, and the first law states that $dU = T \, dS + H \, dM$. The equation of state is given by Curie's law, $M = \kappa H/T$, where $\kappa > 0$ is a constant. This is valid only at small $H$ and high $T$. The heat capacity at constant $H$ is denoted by $C_H$, and that at constant $M$ is $C_M$. Many thermodynamic relations can be obtained from those of a $P$–$V$–$T$ system by the correspondence $H \leftrightarrow -P$, $M \leftrightarrow V$.

(a) Show that

$$C_M = \left(\frac{\partial U}{\partial T}\right)_M, \qquad C_H = \left(\frac{\partial U}{\partial T}\right)_H - H\left(\frac{\partial M}{\partial T}\right)_H.$$

(b)  From the analog of the energy equation and Curie's law, show that

$$\left(\frac{\partial U}{\partial M}\right)_T = 0.$$

This is the analog of the statement that the internal energy of the ideal gas is independent of the volume.

(c)  Show that

$$C_H - C_M = \frac{M^2}{\kappa}.$$

**3.9**  Define the free energy $A(M, T)$ and Gibbs potential $G(H, T)$ of a magnetic system by analogy with the $P$–$V$–$T$ system.

(a)  Show that

$$dA = -S\,dT + H\,dM,$$
$$dG = -S\,dT - M\,dH.$$

(b)  Show that $(\partial S/\partial H)_T = (\partial M/\partial T)_H$; hence

$$\left(\frac{\partial S}{\partial H}\right)_T = -\frac{\kappa H}{T^2}.$$

(c)  With the help of the chain rule, show that

$$\left(\frac{\partial T}{\partial H}\right)_S = \frac{\kappa H}{C_H T}.$$

**3.10**  *Adiabatic demagnetization:* A paramagnet cools when the magnetic field is decreased adiabatically. The path shown in the accompanying diagram can be used to cool the system to very low temperatures.

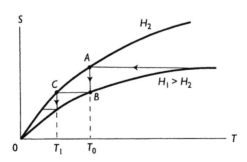

(a)  Using the properties derived in the last problem, verify the qualitative behavior of $S$ as a function of $T$ and $H$ .

(b) Put $H_2 = 0$, $H_1 = H$. Assume $C_M = aT^3$, where $a$ is a constant. Find the heat absorbed by the system along the path of isothermal magnetization $A \rightarrow B$.

(c) In the adiabatic demagnetization $B \rightarrow C$, calculate $T_1$ as a function of $T_0$ and $H$.

# Chapter 4

# Phase transitions

## 4.1 First-order phase transition

When water boils, it undergoes a phase transition from a liquid to a gas phase. The equation of state in each of these phases is a regular function, continuous with continuous derivatives; but in going from one phase to another it abruptly changes to a different regular function. The liquid–gas transition is a "first-order phase transition", in that the first derivative of the Gibbs potential is discontinuous across the phase boundary. By the Maxwell relations $V = -(\partial G/\partial P)_T$ and $S = -(\partial G/\partial T)_P$, the density and entropy are discontinuous.

There are second-order phase transitions, in which the first derivatives of $G$ are continuous, but the second derivatives change discontinuously. Examples include the gas–liquid transition at the critical point, the ferromagnetic transition, and the superconducting transition. We shall discuss the first-order transition in more detail.

An isotherm exhibiting a first-order gas–liquid transition is shown in the $P$–$V$ diagram of Fig. 4.1. At the point 1 the system is all liquid, at point 2 it is all gas, and in between the system is a mixture of liquid and gas in the states 1 and 2, respectively. They coexist in thermal equilibrium, at a pressure called the *vapor pressure*. Since the two phases have different densities, the total volume changes at constant pressure as the relative proportion is changed, generating the horizontal portion of the isotherm.

The $P$–$V$ diagram is a projection of the equation-of-state surface shown in Fig. 4.2, where we have included a first-order liquid–solid phase transition as well. The gas–liquid transition region is topped by a critical point, which lies on the critical isotherm. The transition regions are ruled surfaces perpendicular to the pressure axis. When projected onto the $P$–$T$ plane, they appear as lines representing phase boundaries. This is depicted in Fig. 4.3.

Following are some important properties:

- *Latent heat*: Since the coexisting phases have different entropies, the system must absorb or release heat during a phase transition. When a unit amount of phase 1 is converted into phase 2, an amount of *latent heat* is liberated:

$$l = T_0(s_2 - s_1), \tag{4.1}$$

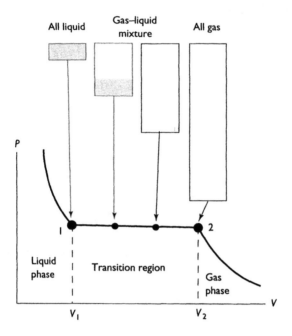

Figure 4.1 An Isotherm exhibiting a first-order phase transition.

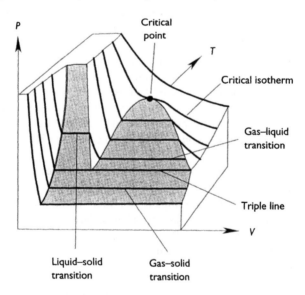

Figure 4.2 Equation of state surface, showing gas–liquid and liquid–solid phase transitions, both of first order. (Not to scale.) Isotherms are shown as heavy lines, and the transition regions are shaded gray.

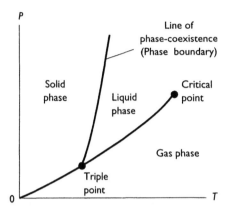

P

Solid
phase

Line of
phase-coexistence
(Phase boundary)

Liquid
phase

Critical
point

Gas phase

Triple
point

0

T

*Figure 4.3* The *P–T* diagram shows phase boundaries, or lines of phase coexistence.

where $T_0$ is the absolute temperature at which the two phases can coexist, and $s_i$ is the specific entropy of the $i$th phase. The specific entropy may mean entropy per particle, per mole, per unit mass, or per unit volume.

- *Critical point*: At the *critical point* the gas and the liquid have equal density and specific entropy. The phase transition becomes second-order at this point. There is no critical point for the liquid–solid transition.
- *Triple point*: The line along which gas, liquid, and solid can all coexist projects onto the *triple point* in the *P–T* diagram.

## 4.2   Condition for phase coexistence

In a first-order transition, the coexisting phases have the same $P$ and $T$. The condition for equilibrium, therefore, is that the total Gibbs potential be minimum. The relative proportion of the two phases present may vary, but each phase is characterized by its chemical potential – the Gibbs potential per particle. We use the equivalent quantity $g_i(P, T)$, the Gibbs potential per unit mass, and represent the total Gibbs potential in the form

$$G = m_1 g_1 + m_2 g_2, \tag{4.2}$$

where $m_i$ is the mass of phase $i$, with a fixed total mass $m = m_1 + m_2$. The contributions from the two phases are additive because we have neglected surface effects. In equilibrium, $G$ is at a minimum with respect to mass transfer from one phase to the other. If we transfer a small amount of mass $\delta m$ from one phase to the other at constant $P$ and $T$, the change $\delta G$ should vanish, to first order:

$$\delta G = g_1 \delta m_1 + g_2 \delta m_2 = 0, \tag{4.3}$$

with the constraint $\delta m_1 = -\delta m_2 = \delta m$. Therefore, the condition for phase coexistence is $(g_1 - g_2)\delta m = 0$. Since $\delta m$ is arbitrary, we must have

$$\Delta g \equiv g_2 - g_1 = 0. \tag{4.4}$$

That is, the coexisting phases must have equal chemical potential. The derivatives of $g$ are discontinuous across the phase boundary:

$$\left(\frac{\partial g}{\partial T}\right)_P = -s \quad \text{(specific entropy).}$$
$$\left(\frac{\partial g}{\partial P}\right)_T = v \quad \text{(specific volume).} \tag{4.5}$$

As the name first-order transition implies, there are finite differences in the specific entropy and specific volume:

$$\Delta s \equiv s_2 - s_1 = -\frac{\partial(\Delta g)}{\partial T} > 0,$$
$$\Delta v \equiv v_2 - v_1 = \frac{\partial(\Delta g)}{\partial P} > 0. \tag{4.6}$$

## 4.3  Clapeyron equation

Since the three variables $\Delta g, \Delta s, \Delta v$ are functions of $P, T$, there must exist a relation among them $f(\Delta g, \Delta s, \Delta v) = 0$, and we can apply the chain rule in the form

$$\left(\frac{\partial(\Delta g)}{\partial T}\right)_P \left(\frac{\partial T}{\partial P}\right)_{\Delta g} \left(\frac{\partial P}{\partial(\Delta g)}\right)_T = -1, \tag{4.7}$$

or

$$\frac{(\partial(\Delta g)/\partial T)_P}{(\partial(\Delta g)/\partial P)_T} = -\left(\frac{\partial P}{\partial T}\right)_{\Delta g}. \tag{4.8}$$

In equilibrium, $\Delta g = 0$, and $(\partial P/\partial T)_{\Delta g=0}$ gives the slope of the transition line on the $P$–$T$ diagram:

$$\frac{dP}{dT} \equiv \left(\frac{\partial P}{\partial T}\right)_{\Delta g=0}. \tag{4.9}$$

For the gas–liquid transition, $P$ is the *vapor pressure*. Substituting this definition into (4.7), and using (4.6), we obtain

$$\frac{dP}{dT} = \frac{\Delta s}{\Delta v}, \tag{4.10}$$

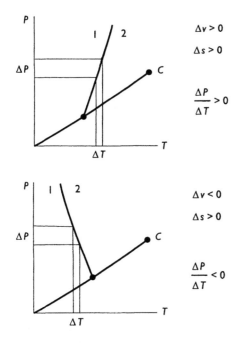

$\Delta v > 0$

$\Delta s > 0$

$$\frac{\Delta P}{\Delta T} > 0$$

$\Delta v < 0$

$\Delta s > 0$

$$\frac{\Delta P}{\Delta T} < 0$$

*Figure 4.4* The slope of the liquid–solid boundary can have either sign, depending on whether the liquid contracts or expands upon freezing.

or

$$\frac{\mathrm{d}P}{\mathrm{d}T} = \frac{l}{T \Delta v},$$

(4.11)

where $l = T \Delta s$ is the latent heat. This is called the *Clapeyron equation*. Depending on the sign of $\Delta v$, the slope $\mathrm{d}P/\mathrm{d}T$ may be positive or negative, as illustrated in Fig. 4.4. The upper panel shows the $P$–$T$ diagram for a substance like $CO_2$, which contracts upon freezing. The lower panel shows that for $H_2O$, which expands upon freezing.

In a second-order phase transition the first derivatives of $g$ vanish, and the Clapeyron equation is replaced by a condition involving second derivatives. (See Problem 4.10.)

## 4.4   van der Waals equation of state

The gas–liquid phase transition owes its existence to intermolecular interactions. The potential energy $U(r)$ between two molecules as a function of their separation $r$ have the qualitative form shown in Fig. 4.5. It has a repulsive core whose radius $r_0$ is of the order of a few angstroms, due to the electrostatic repulsion between the

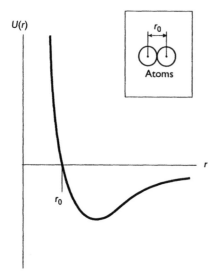

*Figure 4.5* Intermolecular potential.

electron clouds. The inset in Fig. 4.5 shows why $r_0$ can be regarded as the molecular diameter. Outside the repulsive core is an attractive tail due to mutual electrostatic polarization. The depth of the attractive part is generally of the order of 1 eV, but varies widely among molecular species. It gives rise to chemical valence and the crystal structure of solids, and thus plays an important role in the everyday world.

To take the intermolecular interaction into account in a qualitative fashion, we separate the effects of the repulsive core and the attractive tail. The hard core excludes a certain volume around a molecule, so other molecules have less room in which to move. The effective volume is therefore smaller than the actual volume:

$$V_{\text{eff}} = V - b, \tag{4.12}$$

where $V$ is the total volume of the system, and $b$ is the total excluded volume, of the order of $b \approx N\pi r_0^3/6$. For a fixed number of atoms, it is a constant parameter.

The pressure of a gas arises from molecules striking the wall of the container. Compared with the case of the ideal gas, a molecule in a real gas hits the wall with less kinetic energy, because it is being held back by the attraction of neighboring molecules. The reduction in the pressure is proportional to the number of pairs of interacting molecules near the wall, and thus to the density. Accordingly we put

$$P = P_{\text{kinetic}} - \frac{a}{V^2}, \tag{4.13}$$

where $P_{\text{kinetic}}$ is the would-be pressure in the absence of attraction, and $a$ is a constant proportional to $N^2$. Van der Waals makes the assumption that, for 1 mol of gas,

$$P_{\text{kinetic}} V_{\text{eff}} = RT. \tag{4.14}$$

This leads to

$$(V - b)\left(P + \frac{a}{V^2}\right) = RT, \tag{4.15}$$

the *van der Waals equation of state*. In this simple model, the substance is characterized by only two parameters $a$ and $b$.

## 4.5 Virial expansion

The van der Waals equation of state approaches that of the ideal gas in the low density limit: $V \to \infty$ at fixed $N$. The successive corrections to the ideal gas equation can be obtained by expanding in powers of $V^{-1}$. We first solve for the pressure:

$$\left(P + \frac{a}{V^2}\right) = \frac{RT}{(V - b)},$$

$$P = \frac{RT}{V}\left(1 - \frac{b}{V}\right)^{-1} + \frac{a}{V^2},$$

$$\frac{PV}{RT} = \left(1 - \frac{b}{V}\right)^{-1} - \frac{a}{RTV}. \tag{4.16}$$

Expanding the right-hand side, we obtain

$$\frac{PV}{RT} = 1 + \frac{1}{V}\left(b - \frac{a}{RT}\right) + \left(\frac{b}{V}\right)^2 + \left(\frac{b}{V}\right)^3 + \cdots. \tag{4.17}$$

This is of the form of a *virial expansion*:

$$\frac{PV}{RT} = 1 + \frac{c_2}{V} + \frac{c_3}{V^2} + \cdots, \tag{4.18}$$

where $c_n$ is called the $n$th *virial coefficient*. The *second virial coefficient*

$$c_2 = b - \frac{a}{RT} \tag{4.19}$$

can be obtained experimentally by observing deviations from the ideal gas law. By measuring this as a function of $T$, we can extract the molecular parameters $a$ and $b$.

## 4.6 Critical point

The van der Waals isotherms are sketched in Fig. 4.6. The equation is a cubic polynomial in $V$:

$$(V - b)\left(PV^2 + a\right) = RTV^2,$$
$$PV^3 - (bP + RT)V^2 + aV - ba = 0. \tag{4.20}$$

There is a region in which the polynomial has three real roots. As we increase $T$ these roots move closer together, and merge at $T = T_c$, the *critical point*. For $T > T_c$, one real root remains, while the other two become a complex-conjugate pair. We can find the critical parameters $P_c$, $V_c$, $T_c$, as follows. At the critical point the equation of state must be of the form

$$(V - V_c)^3 = 0,$$
$$V^3 - 3V_c V^2 + 3V_c^2 V - V_c^3 = 0. \tag{4.21}$$

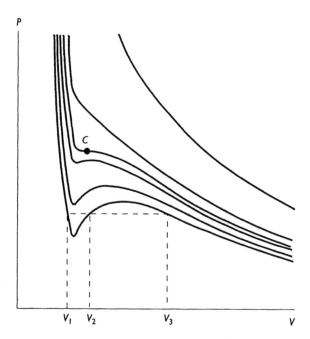

Figure 4.6 Isotherms of the van der Waals gas. The critical point is at C.

Comparison with (4.20) yields

$$3V_c = b + \frac{RT_c}{P_c},$$

$$3V_c^2 = \frac{a}{P_c},$$

$$V_c^3 = \frac{ba}{P_c}. \tag{4.22}$$

They can be solved to give

$$RT_c = \frac{8a}{27b},$$

$$P_c = \frac{a}{27b^2},$$

$$V_c = 3b. \tag{4.23}$$

## 4.7  Maxwell construction

The van der Waals isotherm is a monotonic function of $V$ for $T > T_c$. Below $T_c$, however, there is a "kink" exhibiting negative compressibility. This is unphysical, and its origin can be traced to the implicit assumption that the density is always uniform. Actually, as we shall see, the system prefers to undergo a first-order phase transition, by breaking up into a mixture of phases of different densities.

According to the Maxwell relation $P = -(\partial A / \partial V)_T$ , the free energy can be obtained as the area under the isotherm:

$$A(V, T) = -\int_{\text{isotherm}} P \, dV. \tag{4.24}$$

Let us carry out the integration graphically, as indicated in Fig. 4.7. The volumes $V_1, V_2$ are defined by the double-tangent construction. At any point along the tangent, such as X, the free energy is a linear combination of those at 1 and 2, and thus represents a mixture of two phases. This non-uniform state has the same $P$ and $T$ as the uniform state 3, but it has a lower free energy, as is obvious from the graphical construction. Therefore the phase-separated state is the equilibrium state.

The states 1 and 2 are defined by the conditions

$$\frac{\partial A}{\partial V_1} = \frac{\partial A}{\partial V_2} \quad \text{(equal pressure)}, \tag{4.25}$$

$$\frac{A_2 - A_1}{V_2 - V_1} = \frac{\partial A}{\partial V_1} \quad \text{(common tangent)}. \tag{4.26}$$

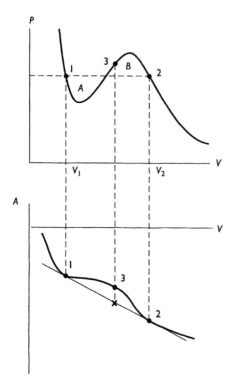

*Figure 4.7* The Maxwell construction.

Thus,

$$- (A_2 - A_1) = -\frac{\partial A}{\partial V_1}(V_2 - V_1),$$  (4.27)

$$\int_{V_1}^{V_2} P\,dV = P_1(V_2 - V_1).$$  (4.28)

This means the areas $A$ and $B$ in Fig. 4.7 are equal to each other. The horizontal line in Fig. 4.7 is known as the *Maxwell construction*.

## 4.8  Scaling

The van der Waals equation of state assumes a universal form if we measure $P, V$ and $T$ in terms of their critical values. Introducing the dimensionless quantities

$$\bar{P} = \frac{P}{P_c}, \quad \bar{T} = \frac{T}{T_c}, \quad \bar{V} = \frac{V}{V_c},$$  (4.29)

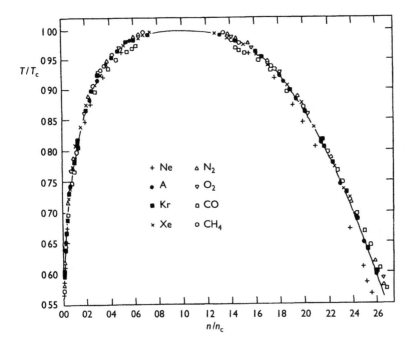

*Figure 4.8* The boundary curve of the gas–liquid transition region becomes universal when the temperature $T$ and the density $n$ are measured in terms of their values at the critical point. After Guggenheim (1945).

we can rewrite the equation of state in the form

$$\left(\bar{V} - \frac{1}{3}\right)\left(\bar{P} + \frac{3}{\bar{V}^2}\right) = \frac{8}{3}\bar{T}. \tag{4.30}$$

This is a universal equation, for the coefficients $a$ and $b$ have disappeared.

The scaling law is partly borne out by experiments. The critical data for the elements involved vary widely, as shown in Table 4.1. Yet, the scaled densities of the liquid and gas at the phase boundary depend on the scaled temperature in a universal manner, as shown in Fig. 4.8. The universal behavior can be fit by

$$\frac{n_L - n_G}{n_c} = \frac{7}{2}\left(1 - \frac{T}{T_c}\right)^{1/3}, \tag{4.31}$$

where $n_L$ is the density of the liquid phase and $n_G$, that of the gas phase. The exponent $1/3$ is a "critical exponent" conventionally denoted by the symbol $\beta$. Although the van der Waals equation of state predicts the scaling behavior, it gives $\beta = 1/2$ instead of the experimental value. This indicates that the model is not adequate for quantitative purposes. (Huang 1987, Section 17.5.)

Table 4.1  Critical data

| Element | $T_c$ (°C) | $P_c$ (atm) |
|---|---|---|
| Ne | −288.7 | 26.9 |
| A | −122.3 | 48.0 |
| Kr | −63.8 | 54.3 |
| Xe | 16.6 | 58.0 |
| $N_2$ | −14.6 | 58.0 |
| $O_2$ | −118.4 | 50.1 |
| CO | −140.0 | 34.5 |
| $CH_4$ | −82.1 | 45.8 |

## Problems

**4.1**  Calculate the change in entropy of the "universe" when 10 kg of water, initially at 20°C, is placed in thermal contact with a heat reservoir at −10°C, until it becomes ice at that temperature. Assume that the entire process takes place under constant atmospheric pressure. The following data are given:

$C_P$ of water = 4,180 J/kg deg,
$C_P$ of ice = 2,090 J/kg deg,
Heat of fusion for ice = $3.34\times10^5$ J/kg.

**4.2**  Integrate the Clapeyron equations near a triple point to obtain the equations for the three transition lines meeting at that point. Make the following assumptions:

- the gas can be treated as ideal,
- the latent heats are constants,
- the molar volumes of solid and liquid are nearly equal constants, negligible compared to that of gas.

**4.3**  Water expands upon freezing, so $dP/dT < 0$ on the P–T diagram (see Fig. 4.4). Calculate the rate of change of the melting temperature of ice with respect to pressure from the Clapeyron equation, using the following data:

Heat of melting of ice  $\ell = 1.44$ J
Molar volume of ice  $v_1 = 20$ cm$^3$
Molar volume of water  $v_2 = 18$ cm$^3$

**4.4**  Reconsider the model of a liquid in Problem 3.5, with free energy

$$A(V,T) = Na_0(T)[v_0(T) - v]^2 - Nf(T),$$

where $v = V/N$, and $v_0(T), a_0(T), f(T)$ are functions of the temperature only.

(a) If $v > v_0(T)$, then the pressure would be negative, but this assumes that the liquid is being stretched uniformly. Actually, the liquid would rather not fill the entire volume available, but remain at specific volume $v_0$. Thus, the pressure would remain zero for $v > v_0$. Show this using an argument similar to that for the Maxwell construction. (See accompanying sketch.)

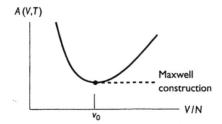

(b) Suppose the liquid is in thermal equilibrium with its vapor, which can be treated as an ideal gas of density $n$. There are two conditions to be fulfilled, the pressures and chemical potentials must equalize. Write down these conditions. Show $v < v_0$, and $v \to v_0$ only at $T = 0$.

(c) For $nkT/a_0 \ll 1$, show $n\lambda^3 = e^{-f/kT}$, where $\lambda = (2\pi\hbar^2/mkT)^{1/2}$.

**4.5** (a) In a liquid–gas transition, the specific volume of the liquid (phase 1) is usually negligible compared with that of the gas (phase 2), which usually can be treated as an ideal gas. Let $\ell = T(s_2 - s_1)$ be the latent heat of evaporation per particle. Under the approximations mentioned, show that

$$\frac{T}{P}\frac{dP}{dT} = \frac{\ell}{kT}.$$

(b) Use this formula to obtain the latent heat per unit mass for liquid $^3$He, in 0.2 K increments of $T$, from the following table of vapor pressures of $^3$He. The mass of a $^3$He atom is $m = 5.007 \times 10^{24}$ g.

| T (K) | P (microns of Hg) |
|---|---|
| 0.200 | 0.0121 |
| 0.201 | 0.0130 |
| 0.400 | 28.12 |
| 0.401 | 28.71 |
| 0.600 | 544.5 |
| 0.601 | 550.3 |
| 0.800 | 2892 |
| 0.801 | 2912 |
| 1.000 | 8842 |
| 1.005 | 9053 |
| 1.200 | 20163 |
| 1.205 | 20529 |

**4.6**   Sketch the Gibbs potential $G(P, T)$ of the van der Waals gas as a function of $P$ at constant $T$. In particular, show the behavior in the transition region. Derive the Maxwell construction using the principle of minimization of $G$.

**4.7** (a)   Calculate the free energy $A(V, T)$ for 1 mol of a van der Waals gas. (*Hint:* Integrate $-\int P\,dV$ along an isotherm. Determine the unknown additive function of $T$ by requiring $A(V, T)$ to approach that of an ideal gas as $V \to \infty$, given in Problem 3.2.)

(b)   Show that $C_V$ of a van der Waals gas is a function of $T$ only. (*Hint:* Use $C_V = -T(\partial^2 A/\partial T^2)_V$ (Problem 3.4). Show $(\partial C_V/\partial V)_T = 0$.)

**4.8**   Find the relationship between $T$ and $V$ for the reversible adiabatic transformation of a van der Waals gas, assuming that $C_V$ is a constant. Check that it reduces to the ideal gas result when $a = b = 0$. (*Hint:* Use the $T\,dS$ equation in (3.16).)

**4.9**   Form the attached table of the second virial coefficient for 1 mol of Ne, find the best fits for the coefficients $a$ and $b$ in the van der Waals equation of state (Data from Holborn 1925).

| $T$ (K) | $c_2$ (cm³ mol⁻¹) |
|---|---|
| 60 | $-20$ |
| 90 | $-8$ |
| 125 | 0 |
| 175 | 7 |
| 225 | 9 |
| 275 | 11 |
| 375 | 12 |
| 475 | 13 |
| 575 | 14 |
| 675 | 4 |

**4.10**   The transition line of a second-order phase transition is shown in the accompanying $P$–$T$ diagram. The first derivatives of the Gibbs potential, giving specific density and entropy, are continuous across this line. The second derivatives, such as compressibilities, are discontinuous. Show that the slope of the transition line is given by

$$\frac{dP}{dT} = \frac{\alpha_1 - \alpha_2}{\kappa_{T1} - \kappa_{T2}},$$

where $\alpha$ denotes the coefficient of thermal expansion, and $\kappa_T$ the isothermal compressibility.

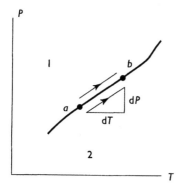

(*Hint*: Calculate the change of volume $\Delta V$ along the transition line. The results for phase 1 and phase 2 must agree.)

# Chapter 5

# The statistical approach

## 5.1 The atomic view

Experiments show that 1 g mol of any dilute gas occupies the same *molar volume*

$$V_0 = 2.24 \times 10^4 \, \text{cm}^3 \qquad (5.1)$$

at STP (standard temperature and pressure):

$$T_0 = 273.15 \, \text{K},$$
$$P_0 = 1 \, \text{atm}. \qquad (5.2)$$

From this we can obtain the gas constant $R = P_0 V_0 / T_0$. The ratio of the molar volume to Avogadro's number gives the density of any gas at STP:

$$\text{Density} = 2.70 \times 10^{19} \, \text{atoms/cm}^3. \qquad (5.3)$$

This indicates the large number of atoms present in a macroscopic volume. We use the term "atom" here in a generic sense, to denote the smallest unit in the gas, which may in fact be a diatomic molecule such as $H_2$.

A gas can approach thermal equilibrium because of atomic collisions. The scattering cross section between atoms is of the order of

$$\sigma = \pi r_0^2,$$
$$r_0 \approx 10^{-8} \, \text{cm}, \qquad (5.4)$$

where $r_0$ is the effective atomic diameter. The average distance traveled by an atom between two successive collisions is called the *mean free path*, whose order of magnitude is given by

$$\lambda \approx \frac{1}{n\sigma}, \qquad (5.5)$$

where $n$ is the particle density. This can be seen as follows. Consider a cylindrical volume inside the gas, of cross-sectional area $A$, as depicted in Fig. 5.1. If the

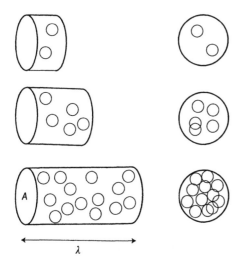

*Figure 5.1* The atoms in the project view begin to overlap when the length of the cylinder is of the order of the mean free path λ.

length of the cylinder is small, it contains few atoms, and in the projected end view the images of the atoms rarely overlap. As the length of the cylinder increases, more and more atoms are included, and the images begin to overlap when the length of the cylinder is of the order of the mean free path. At that point the area $A$ is roughly equal to the cross-sectional area of all the atoms: $A = (A\lambda)(n\sigma)$, which gives (5.5). At STP we have

$$\lambda \approx 10^{-5} \text{ cm}. \tag{5.6}$$

Thus, on a macroscopic scale, an atom can hardly move without colliding with another atom.

We have learnt from thermodynamics that the internal energy per particle in an ideal gas is $\frac{3}{2}kT$. Equating this to the average kinetic energy of an atom, we obtain

$$\frac{1}{2}mv^2 = \frac{3}{2}kT. \tag{5.7}$$

Thus, the average velocity is

$$v = \sqrt{\frac{3kT}{m}}. \tag{5.8}$$

For numerical calculations, it is useful to remember that

$$kT \approx \frac{1}{40} \text{ eV} \quad \text{at } T = 300 \text{ K},$$

$$m_e c^2 \approx 0.5 \text{ MeV},$$

(5.9)

where $m_e$ is the mass of the electron, and $c$ is the velocity of light. Thus, for $H_2$ gas at 300 K, we have

$$\frac{v^2}{c^2} = \frac{3kT}{mc^2} \approx 3 \left( \frac{1}{40} \text{ eV} \right) \frac{1}{0.5 \times 10^6 \times 2 \times 2000 \text{ eV}} \approx 4 \times 10^{-11}. \quad (5.10)$$

Using, $c = 3 \times 10^{10}$ m/s, we get

$$v \approx 10^5 \text{ cm/s}. \tag{5.11}$$

This is the speed with which a gas expands into a vacuum in a free expansion. The average time between successive collisions is called the *collision time*:

$$\tau = \frac{\lambda}{v} \approx 10^{-10} \text{ s}, \tag{5.12}$$

which is of the order of the relaxation time towards local thermal equilibrium.

## 5.2  Phase space

In classical mechanics, the state of an atom at any instant of time is specified by its position $\mathbf{r}$ and momentum $\mathbf{p}$. The six components of these quantities span the phase space of one atom. For $N$ atoms, the total number of degrees of freedom is $6N$, and the total phase space is a $6N$-dimensional space. The motions of the particles are governed by the Hamiltonian

$$H(p, r) = \sum_{i=1}^{N} \frac{\mathbf{p}_i^2}{2m} + \frac{1}{2} \sum_{i \neq j} U(\mathbf{r}_i - \mathbf{r}_j) \tag{5.13}$$

where $U(\mathbf{r})$ is the interatomic potential. The Hamiltonian equations of motions are

$$\dot{\mathbf{p}}_i = \frac{\partial H}{\partial \mathbf{r}_i},$$

$$\dot{\mathbf{r}}_i = -\frac{\partial H}{\partial \mathbf{p}_i}.$$

(5.14)

In the absence of external time-dependent forces, $H$ has no explicit dependence on the time. The value of the Hamiltonian is the energy, which is a constant of the motion.

We use the shorthand $(p, r)$ to denote all the momenta and coordinates. The $6N$-dimensional space spanned by $(p, r)$ is called the $\Gamma$-space. A point in this space, called a *representative point*, corresponds to a state of the $N$-body system at a particular time. As time evolves, the representative point traces out a trajectory. It never intersects itself, because the solution to the equations of motion is unique, given initial conditions. Because of energy conservation, it always lies on an *energy surface*, a hypersurface in $\Gamma$-space defined by

$$H(p, r) = E. \tag{5.15}$$

Because of atomic collisions, the trajectory is jagged, and exceedingly sensitive to initial conditions. It essentially makes a self-avoiding random walk on the energy surface. A symbolic representation of $\Gamma$-space is indicated in Fig. 5.2.

We can alternatively represent the state of the atomic system by specifying the states of each atom, in the 6D phase space of a single atom spanned by $(\mathbf{p}, \mathbf{r})$. This space is called the $\mu$-space, and the system is represented by $N \approx 10^{19}$ points, as illustrated schematically in Fig. 5.2. As time evolves, these points move and collide with one another, and the aggregate billows like a cloud.

We shall confine our discussion to the ideal gas, which is the low-density limit of a real gas. In this limit the atomic interactions make only small contributions to the total energy, and we can approximate the total energy by the sum of energies of the individual atoms. However, the interactions cannot be entirely ignored, for they are responsible for the establishment of thermal equilibrium. It is important to realize that the ideal gas is not a gas with no interactions, but corresponds to the limiting case in which the interaction potential approaches zero.

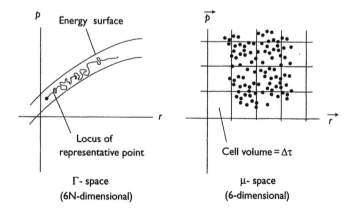

*Figure 5.2* In $\Gamma$-space, the history of an $N$-particle system is represented by one trajectory on a $(6N - 1)$-dimensional energy surface. In $\mu$-space, it is represented by a cloud of $N$ points in a 6D space.

## 5.3   Distribution function

We are interested in the statistical properties of the atomic system, rather than the behavior of the individual atoms. In particular, we wish to calculate the distribution function $f(\mathbf{p}, \mathbf{r}, t)$ defined as follows. Divide $\mu$-space into cells, which are 6D volume elements

$$\Delta \tau = \Delta p_x \, \Delta p_y \, \Delta p_z \, \Delta x \, \Delta y \, \Delta z \tag{5.16}$$

A cell is assumed to be large enough to contain a large number of atoms and yet small enough to be considered infinitesimal on a macroscopic scale. From a macroscopic point of view, atoms in the $i$th cell have unresolved positions $\mathbf{r}_i$, and momenta $\mathbf{p}_i$, and a common energy $\epsilon_i = \mathbf{p}_i^2/2m$. The number of atoms in cell $i$ at time $t$ is called the *occupation number* $n_i$, and the distribution function is the occupation number per unit volume:

$$f(\mathbf{p}_i, \mathbf{r}_i, t)\Delta \tau = n_i. \tag{5.17}$$

Since there are $N$ atoms with total energy $E$, we have the conditions

$$\sum_i n_i = N,$$
$$\sum_i n_i \epsilon_i = E. \tag{5.18}$$

The unit for the phase–space volume $\Delta\tau$ is so far arbitrary. This does not lead to ambiguities when $f \Delta\tau$ appear together, but when $f$ appears alone, as in the expression for entropy later, we will have an arbitrary undetermined constant. As we shall see, quantum mechanics ultimately determines the unit to be $h^3$, where $h$ is Planck's constant.

In the thermodynamic limit, we assume that $f(\mathbf{p}, \mathbf{r}, t)$ approaches a continuous function, and regard $\Delta\tau$ as mathematically infinitesimal:

$$\Delta \tau = \mathrm{d}^3p \, \mathrm{d}^3r. \tag{5.19}$$

We can then write

$$\int \mathrm{d}^3p \, \mathrm{d}^3r f(\mathbf{p}, \mathbf{r}, t) = N,$$
$$\int \mathrm{d}^3p \, \mathrm{d}^3r f(\mathbf{p}, \mathbf{r}, t)\frac{\mathbf{p}^2}{2m} = E. \tag{5.20}$$

If the density is uniform, then $f(\mathbf{p}, \mathbf{r}, t)$ is independent of $\mathbf{r}$. We denote it by $f(\mathbf{p}, t)$, and write

$$\int \mathrm{d}^3p f(\mathbf{p}, t) = \frac{N}{V},$$
$$\int \mathrm{d}^3p \, f(\mathbf{p}, t)\frac{\mathbf{p}^2}{2m} = \frac{E}{V}. \tag{5.21}$$

The distribution function evolves in time according to the microscopic equations of motion, and we assume that it will eventually approach a time-independent form $f_0(\mathbf{p}, \mathbf{r})$, which corresponds to thermal equilibrium. Our task is to find the equilibrium distribution, and to deduce from it the thermodynamics of the ideal gas.

## 5.4  Ergodic hypothesis

The representative point in $\Gamma$-space, corresponds to a state of the $N$-particle system, and traces out a trajectory on an energy surface as time goes on. After some time, long compared to the collision time, about $10^{-10}$ s for a gas at STP, the trajectory should reach some kind of steady state, which corresponds to thermal equilibrium.

We know from mechanics that, for $N > 3$, the motion is generally "chaotic", in the sense that two points initially close to each other will diverge exponentially with time. At successive time intervals which are large compared to the collision time, we expect that the trajectory in $\Gamma$-space performs a random walk over the whole energy surface. This is the essence of the *ergodic hypothesis*. More precisely, one can prove the following for certain artificial systems:

> Given a sufficiently long time, the representative point of an isolated system will come arbitrarily close to any given point on the energy surface.

Ergodic theory is a rigorous mathematical discipline that aims to prove statements like the above for various choices of systems, but it does not provide a criterion for "sufficiently long time". The question of relaxation time is a dynamical one, while the methods of ergodic theory are based on avoidance of dynamics. Thus, although the ergodic theorem gives us conceptual understanding, it does not hold any promise for practical applications.

## 5.5  Statistical ensemble

Our measuring instruments have finite resolutions, and they effectively perform averages over time intervals that are small on a macroscopic scale, but large compared to the collision time. The basis of the statistical method is the assumption that this time averaging can be replaced by an average over a suitably chosen collection of systems called a *statistical ensemble*. This is an infinite collection of identical copies of the system, characterized by a distribution function $\rho(p, r, t)$ in $\Gamma$-space:

$$\rho(p, r, t)\mathrm{d}p\,\mathrm{d}r = \text{Number of systems in } \mathrm{d}p\,\mathrm{d}r \text{ at time } t; \qquad (5.22)$$

where

$$(p, r) = (\mathbf{p}_1, \dots, \mathbf{p}_N \,;\, \mathbf{r}_1, \dots, \mathbf{r}_N),$$
$$\mathrm{d}p\,\mathrm{d}r = \mathrm{d}^{3N}p\,\mathrm{d}^{3N}r. \qquad (5.23)$$

The probability, per unit phase–space volume, of finding the system in $dp\,dr$ at time $t$ is given by

$$\text{Probability density} = \frac{\rho(p,r,t)}{\int dp\,dr\,\rho(p,r,t)}. \tag{5.24}$$

The *ensemble average* of a physical quantity $O(p, r)$ is defined as

$$\langle O \rangle = \frac{\int dp\,dr\,\rho(p,r,t)\,O(p,r)}{\int dp\,dr\,\rho(p,r,t)}. \tag{5.25}$$

It is important to keep in mind that members of the ensemble are mental copies of the system, and do not interact with one another.

As the system approaches thermal equilibrium, the ensemble evolves into an equilibrium ensemble with a time-independent distribution function $\rho(p,r)$. The ensemble average with respect to $\rho(p, r)$ then yields thermodynamic quantities. We assume that $\rho(p, r)$ depends on $(p, r)$ only through the Hamiltonian, and denote it by $\rho(H(p,r))$. This automatically makes it time-independent, since the Hamiltonian is a constant of the motion.

## 5.6   Microcanonical ensemble

For an isolated system, $\rho$ is constant over an energy surface, according to the ergodic hypothesis. This condition is known as the assumption of equal a priori probability, and defines the *microcanonical ensemble*:

$$\rho(H(p,r)) = \begin{cases} s1 & \text{if } E < H(p,r) < E + \Delta, \\ 0 & \text{otherwise,} \end{cases} \tag{5.26}$$

where $\Delta$ is some fixed number that specifies the tolerance of energy measurements, with $\Delta \ll E$. The volume occupied by the microcanonical ensemble is, in arbitrary units,

$$\Gamma(E, V) = \int dp\,dr\,\rho\,(H(p,r)) = \int_{E<H(p,r)<E+\Delta} dp\,dr, \tag{5.27}$$

where the dependence on the spatial volume $V$ comes from the limits of the integrations over $dr$. This is the volume of the shell bounded by the two energy surfaces with respective energies $E + \Delta$ and $E$. Since $\Delta \ll E$, it can be obtained by multiplying the surface area of the energy surface $E$ by the thickness $\Delta$ of the shell. The surface area, in turn, can be obtained from the volume of the interior of the

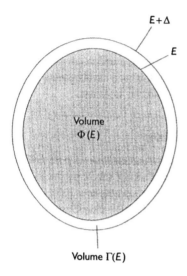

*Figure 5.3* The volume $\Gamma$ of the thin shell can be obtained from the volume $\Phi$ of the interior of the shell: $\Gamma = (\partial\Phi/\partial E)\Delta$.

surface. Thus,

$$\Gamma(E, V) = \frac{\partial\Phi(E, V)}{\partial E}\Delta,$$

$$\Phi(E, V) = \int_{H(p.q)<E} dp\, dr,$$

$$(5.28)$$

where $\Phi(E, V)$ is the volume of phase space enclosed by the energy surface $E$, as illustrated in Fig. 5.3. Ensemble averages are independent of the unit used for $\Gamma$.

The entropy is defined to be proportional to the logarithm of the number of available states, which is measured by the phase–space volume in $\Gamma$-space:

$$S(E, V) = k \ln \Gamma(E, V).$$

$$(5.29)$$

The factor $k$, which is Boltzmann's constant, establishes the unit. The arbitrariness of the units for $\Gamma$ means that $S$ is defined up to an arbitrary additive constant. The reason for taking the logarithm of $\Gamma$ is to make $S$ additive for independent systems. Two independent non-interacting systems have separate distributions $\rho$ and $\rho'$ respectively, and they occupy volumes $\Gamma$ and $\Gamma'$ in their respective $\Gamma$-spaces. The total $\Gamma$-space is the direct product of the two spaces, and the total volume is the product $\Gamma\Gamma'$. Another way of stating this is that the probability of two independent events is the product of the individual probabilities. Thus, the total entropy is

$$k \ln(\Gamma\Gamma') = k \ln \Gamma + k \ln \Gamma'. \tag{5.30}$$

To obtain thermodynamic functions through the Maxwell relations, we need to obtain the function $E(S, V)$ from $S(E, V)$. Since this is a cumbersome procedure in general, we use simpler alternative routes whenever possible.

## 5.7 The most probable distribution

As we have seen, the state of a gas of $N$ atoms is represented by one point in $\Gamma$-space, and $N$ points in $\mu$-space. If we permute the positions of the $N$ points in $\mu$-space, we will generate a new state of the gas, which corresponds to a different point in $\Gamma$-space. Such a permutation, however, does not affect the distribution $\{n_i\}$, which counts only the number of points in a given volume element, regardless of which ones reside in the volume element. Therefore, a given distribution $\{n_i\}$ corresponds to a collection of states in $\Gamma$-space; it "occupies" a volume in $\Gamma$-space, which we denote by

$$\Omega\{n_i\} \equiv \Omega\{n_1, n_2, \dots\}. \tag{5.31}$$

The total volume in $\Gamma$-space is

$$\Gamma(E, V) = \sum_{\{n_i\}} \Omega\{n_1, n_2, \dots\}, \tag{5.32}$$

where the sum $\sum_{\{n_k\}}$ extends over all possible sets $\{n_k\}$ that satisfy the constraints (5.18). The ensemble average of the occupation number $n_k$ can be written as

$$\langle n_k \rangle = \frac{\sum_{\{n_i\}} n_k \Omega\{n_1, n_2, \dots\}}{\sum_{\{n_i\}} \Omega\{n_1, n_2, \dots\}}. \tag{5.33}$$

On the other hand, we define the *most probable distribution* $\{\bar{n}_i\}$ as the set of occupation numbers that maximizes $\Omega\{n_i\}$ under the constraints (5.18). We shall calculate $\bar{n}_k$ rather than $\langle n_k \rangle$, since that presents an easier mathematical task. We shall show afterwards that $\bar{n}_k \to \langle n_k \rangle$ in the thermodynamic limit.

To find the most probable distribution, picture the cells of $\mu$-space as an array of boxes with $n_i$ identical balls in the $i$th box, as illustrated in Fig. 5.4. The number of

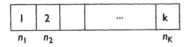

Figure 5.4 Place $n_1$ balls in box 1, $n_2$ balls in box 2, etc.

microscopic states corresponding to this distribution is the number of permutations of the $N$ balls that leaves the distribution unchanged:

$$\Omega\{n_i\} = \frac{N!}{n_1!n_2!\ldots n_K!}g_1^{n_1}\ldots g_K^{n_K},\qquad(5.34)$$

where $g_i$ is the intrinsic probability for the $i$th cell. Actually $g_i = 1$, but it is convenient to keep it arbitrary for mathematical purposes.

Since $\Omega\{\bar{n}_i\}$ is by definition at a maximum, its change upon a variation of the set $\{n_i\}$ must be of second-order smallness. That is, $\delta\Omega\{n_i\} = 0$ under the variation $n_i \to \bar{n}_i + \delta n_i$ with the constraints

$$\sum_i \delta n_i = 0,$$
$$\sum_i \epsilon_i \delta n_i = 0.\qquad(5.35)$$

We use the equivalent condition

$$\delta \ln \Omega\{n_i\} = 0,\qquad(5.36)$$

since the logarithm is a monotonic function.

## 5.8  Lagrange multipliers

Because of the constraints, we cannot vary the $n_i$ independently. Using the method of Lagrange multipliers derived in the Appendix, we consider a modified problem

$$\delta\left[\ln \Omega\{n_i\} + \alpha \sum_i n_i - \beta \sum_i \epsilon_i n_i\right] = 0,\qquad(5.37)$$

where $\alpha$ and $\beta$ are fixed parameters called Lagrange multipliers. We now vary each $n_i$ independently, and after obtaining $\ln \Omega\{\bar{n}_i\}$ as a function of $\alpha$ and $\beta$, we determine $\alpha$ and $\beta$ so as to satisfy the constraints (5.18).

From (5.34) we have

$$\ln \Omega\{n_i\} = \ln N! - \sum_i \ln n_i! + \sum_i n_i \ln g_i\qquad(5.38)$$

Assuming that $N$ and $n_i$ are large compared to unity, we can use the Stirling approximation $\ln n! \approx n \ln n - n$, which is derived in the Appendix, and write

$$\ln N! \approx N \sum_i \ln N - N,$$
$$\sum_i \ln n_i! \approx \sum_i n_i \ln n_i - \sum_i n_i.\qquad(5.39)$$

Setting $\sum_i n_i = N$, we have

$$\ln \Omega\{n_i\} \approx N \ln N - N + \sum_i \left[ -n_i \ln n_i + (1 + \alpha) n_i - \beta \epsilon_i n_i + n_i \ln g_i \right].$$

$$(5.40)$$

Varying the $n_i$ independently, we obtain

$$\sum_i \left[ -\ln n_i + \alpha - \beta \epsilon_i + \ln g_i \right] \delta n_i = 0. \tag{5.41}$$

Since the $\delta n_i$ are arbitrary and independent, we must have

$$\ln n_i = \alpha - \beta \epsilon_i + \ln g_i,$$
$$n_i = g_i C e^{-\beta \epsilon_i}, \tag{5.42}$$

where $C = e^\alpha$. We now set $g_i = 1$ and write

$$n_i = C e^{-\beta \epsilon_i}. \tag{5.43}$$

The Lagrange multipliers are determined by the conditions

$$C \sum_i e^{-\beta \epsilon_i} = N,$$
$$C \sum_i \epsilon_i e^{-\beta \epsilon_i} = E. \tag{5.44}$$

In the limit $N \to \infty$, we make the replacement

$$n_i \to f(\mathbf{p}_i, \mathbf{r}_i) d^3 p d^3 r. \tag{5.45}$$

For a free gas in the absence of external potential, the distribution function is independent of $\mathbf{r}$:

$$f(\mathbf{p}) = C e^{-\beta p^2 / 2m}. \tag{5.46}$$

Thus the conditions become

$$\int d^3 p f(\mathbf{p}) = \frac{N}{V} = n,$$
$$\frac{1}{n} \int d^3 p \frac{\mathbf{p}^2}{2m} f(\mathbf{p}) = \frac{E}{N}. \tag{5.47}$$

The distribution function $f(\mathbf{p})$ is called the *Maxwell–Boltzmann distribution*. It describes a uniform free gas in thermal equilibrium, with density $n$ and energy per particle $E/N$.

## Problems

**5.1** Hydrogen gas is contained in a cylinder at STP. Estimate the number of times the wall of the cylinder is being hit by atoms per second, per unit area.

**5.2** A room of volume $3 \times 3 \times 3 \, m^3$ contains air at STP. Treating the air molecules as independent objects, estimate the probability that you will find a $1 \, cm^3$ volume somewhere in the room totally devoid of air, due to statistical fluctuations. Do the same for a $1 \, Å^3$ volume.

**5.3** In a gas at STP, let $p(r) \, dr$ be the probability that an atom has a nearest neighbor between distances $r$ and $r + dr$. Find $p(r)$.

**5.4** In an atomic beam experiment a collimated beam of neutral $Na$ atoms traverses a vacuum chamber for a distance of 10 cm. How good a vacuum is required for the beam to remain well-collimated during the transit?

**5.5** Neutrinos are the most weakly interacting particles we know of, and they can penetrate the earth with the greatest of ease. This has caused some writer to worry about neutrinos hitting "a lover and his lass" from beneath the bed (by way of Nepal). Is the fear well founded?

(a) Assuming a neutrino cross section of $\sigma = 10^{-40} \, cm^2$ for collision with a nucleon, estimate the neutrino mean free path in water.

(b) Sources for neutrinos include cosmic-ray reactions in the atmosphere, and astrophysical events. For the sake of argument, assume there is a neutrino flux of $50 \, cm^{-2} \, s^{-1}$. Show that a person of mass 150 kg would get hit by a neutrino about once in a lifetime. (Perhaps that's what kills the person.)

**5.6** To calculate the phase–space volume of an ideal gas in the microcanonical ensemble, we need the surface area of an $n$-dimensional sphere of radius $R$. The volume of an $n$-sphere of radius $R$ is of the form

$$\Phi_n(R) = C_n R^n.$$

The surface area is

$$\Sigma_n(R) = n C_n R^{n-1}.$$

All we need is the constant $C_n$. Show that

$$C_n = \frac{2\pi^{n/2}}{\Gamma\left(\frac{n}{2} + 1\right)},$$

where $\Gamma(z)$ is the gamma function.

*Suggestion:* We know that $\int_{-\infty}^{\infty} dx \, \exp\left(-\lambda x^2\right) = \sqrt{\pi/\lambda}$. Thus,

$$\int_{-\infty}^{\infty} dx_1 \ldots dx_n \, e^{-\lambda(x_1^2 + \cdots + x_n^2)} = (\pi/\lambda)^{n/2}.$$

Now rewrite the integral in spherical coordinates as $n C_n \int_0^{\infty} dR \, R^{n-1} \exp\left(-\lambda R^2\right)$, which is a gamma function.

**5.7**   The Hamiltonian for a free classical ideal gas of $N$ atoms can be written as $H = p_1^2 + p_2^2 + \cdots + p_{3N}^2$, where we have chosen units such that $2m = 1$.

(a)  Show that the phase–space volume is $\Gamma(E, V) = K_0 V^N \Sigma_n(\sqrt{E})$, where $K_0$ is a constant, and $n = 3N$.

(b)  Calculate $\Sigma_n$, and obtain the entropy $S(E, V)$. Verify that it agrees with the result from thermodynamics.

**5.8**   If we put the ideal gas in an external harmonic-oscillator potential, the Hamiltonian would become $H = \left(p_1^2 + p_2^2 + \cdots + p_{3N}^2\right) + \left(r_1^2 + r_2^2 + \cdots + r_{3N}^2\right)$, in special units.

(a)  Show that the phase–space volume is $\Gamma(E, V) = K_1 \Sigma_n(\sqrt{E})$, where $K_1$ is a constant, and $n = 6N$.

(b)  Find the entropy of the system.

**5.9**   *Random walk*: An atom in a gas experiences frequent and erratic collisions. A simple way to simulate its motion is to suppose that it executes a random walk. In this problem we study a random walk in one dimension.

Consider a particle making successive random steps along the $x$-axis. The steps are of equal length, and the probability of going forward or backward is $\frac{1}{2}$. Let $W(k, n)$ be the probability that the particle is $k$ steps from the starting point, after taking $n$ random steps. Show the following results:

(a)  The particle made a total of $(n + k)/2$ steps forward, and $(n - k)/2$ steps backward.

(b)  The forward steps can be chosen in $\binom{n}{r}$ ways, where $r = (n + k)/2$. Each of these choices are mutually exclusive, with intrinsic probability $2^{-n}$. Adding these probabilities leads to the result

$$W(k, n) = \frac{2^{-n}}{[(n + k)/2]! \, [(n - k)/2]!}.$$

(c)  For $n \gg 1$, and $n \gg k$, use of the Stirling approximation and the power series expansion for the logarithm leads to

$$W(k, n) \approx \sqrt{\frac{2}{\pi n}} e^{-k^2/2n}.$$

The probability of returning to the origin after a large number of steps $n$ is therefore $W(0, n) \approx \sqrt{2/\pi n}$.

(d)  Define linear distance by $x = k x_0$, where $x_0$ is the size of the step in length units. Define the duration $t$ of a step by $t = n t_0$, where $t_0$ is the duration of a step in seconds. The probability that the particle is located between $x$ and $x \, dx$ at time $t$ is $W(x, t) \, dx$, where

$$W(x, t) = \frac{1}{\sqrt{4\pi D t}} e^{-x^2/4Dt},$$

with

$$D = \frac{x_0^2}{2t_0}.$$

This is the diffusion law, and $D$ is the diffusion constant.

**5.10**   Regard atomic motion in a gas as a random walk due to collisions, give an order-of-magnitude estimate of the time it would take an air molecule in a room to traverse a distance of 1 cm. What about 1 m?

**5.11**   A gas of $N$ atoms can return to an initial state after a certain period of time, through random coincidence. For example, in the free-expansion experiment, the gas escapes into an initially empty compartment. You can expect that the genie will spontaneously return to the bottle after some time.

Based on random walk, give an argument to show that this time is of the order of $e^N$ collision times.

# Chapter 6

# Maxwell–Boltzmann distribution

## 6.1 Determining the parameters

The Maxwell–Boltzmann distribution for an ideal gas is

$$f(\mathbf{p}) = Ce^{-\beta \mathbf{p}^2/2m}. \tag{6.1}$$

The conditions that determine the parameters $C$ and $\beta$ are

$$\int d^3 p f(\mathbf{p}) = n,$$

$$\frac{1}{n} \int d^3 p \, \frac{\mathbf{p}^2}{2m} f(\mathbf{p}) = \frac{E}{N}, \tag{6.2}$$

where $n$ is the density of the gas, and $E/N$ is the energy per particle.

To explicitly calculate $C$ and $\beta$, we need the Gaussian integral

$$\int_{-\infty}^{\infty} dx e^{-\lambda x^2} = \sqrt{\frac{\pi}{\lambda}}. \tag{6.3}$$

Related integrals can be obtained by differentiating both sides with respect to $\lambda$:

$$\int_{-\infty}^{\infty} dx \, x^2 e^{-\lambda x^2} = \frac{\sqrt{\pi}}{2\lambda^{3/2}},$$

$$\int_{-\infty}^{\infty} dx \, x^4 e^{-\lambda x^2} = \frac{3\sqrt{\pi}}{4\lambda^{5/2}}, \tag{6.4}$$

and so on. Thus,

$$n = C \int_{-\infty}^{\infty} dp_1 \, dp_2 \, dp_3 \, e^{-\lambda(p_1^2 + p_2^2 + p_3^2)}$$

$$= C \left[ \int_{-\infty}^{\infty} dp \, e^{-\lambda p^2} \right]^3$$

$$= C \left( \frac{\pi}{\lambda} \right)^{3/2}. \tag{6.5}$$

With $\lambda = \beta/2m$, we obtain

$$C = n \left(\frac{\beta}{2\pi m}\right)^{3/2}.$$ (6.6)

Next we calculate $E/N$, the energy per particle

$$\frac{E}{N} = \frac{C}{2mn} \int_{-\infty}^{\infty} dp_1\, dp_2\, dp_3 \left(p_1^2 + p_2^2 + p_3^2\right) e^{-\lambda(p_1^2 + p_2^2 + p_3^2)}$$

$$= \frac{3}{2m} \left(\frac{\lambda}{\pi}\right)^{3/2} \int_{-\infty}^{\infty} dp_1\, dp_2\, dp_3\, p_1^2\, e^{-\lambda(p_1^2 + p_2^2 + p_3^2)}$$

$$= \frac{3}{2m} \left(\frac{\lambda}{\pi}\right)^{3/2} \left[\int_{-\infty}^{\infty} dp_1\, p_1^2\, e^{-\lambda p_1^2}\right] \left[\int_{-\infty}^{\infty} dp\, e^{-\lambda p^2}\right]^2$$

$$= \frac{3}{2m} \left(\frac{\lambda}{\pi}\right)^{3/2} \frac{\sqrt{\pi}}{2\lambda^{3/2}} \left(\frac{\pi}{\lambda}\right) = \frac{3}{4m\lambda} = \frac{3}{2\beta}.$$ (6.7)

Therefore,

$$\beta = \frac{3}{2} \frac{E}{N}.$$ (6.8)

We shall show that $\beta = 1/kT$.

## 6.2 Pressure of an ideal gas

The pressure of an ideal gas is the average force per unit area that it exerts on the wall of its container. Take the wall to be normal to the $x$-axis, and assume that the wall is perfectly reflecting. When an atom with $x$-component velocity $v_x$ is reflected by the wall, it will transfer an amount of momentum $2mv_x$ to the wall. The force acting on the wall is the momentum transfer per unit time, and the pressure is the force per unit area of the wall:

Pressure = (Momentum transfer per atom) × (Flux of atoms). (6.9)

The flux is the number of atoms crossing unit area per second, and this is equal to the number of atoms contained in a cylinder of length equal to $v_x$, of unit cross section. This is illustrated in Fig. 6.1. Thus,

Flux of atoms = $v_x f(p)\, d^3p$. (6.10)

The pressure is given by

$$P = \int_{v_x > 0} d^3p \,(2mv_x) v_x f(\mathbf{p}) = m \int d^3p\, v_x^2 f(\mathbf{p}).$$ (6.11)

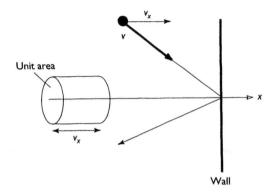

Figure 6.1 In one second, all the atoms with x-component velocity $v_x$ would have evacuated a cylinder of length numerically equal to $v_x$.

In the integrand, we may replace $v_x^2$ by $\frac{1}{3}(v_x^2 + v_y^2 + v_z^2) = \mathbf{p}^2/3m^2$. Thus,

$$P = \frac{1}{3m} \int d^3p\, \mathbf{p}^2 f(\mathbf{p}) = \frac{2}{3} \frac{E}{V}. \qquad (6.12)$$

This gives the equation of state of an ideal gas, and relates the energy per particle to the absolute temperature:

$$\frac{E}{N} = \frac{3}{2}kT. \qquad (6.13)$$

Since we have earlier shown $E/N = 3/(2\beta)$, this gives $\beta = 1/kT$.

The parameters in the Maxwell–Boltzmann distribution are therefore

$$\beta = \frac{1}{kT},$$
$$C = \frac{n}{(2\pi mkT)^{3/2}}. \qquad (6.14)$$

## 6.3 Equipartition of energy

The factor 3 in the formula $E/N = \frac{3}{2}kT$ represents the number of translational degrees of freedom, the number of momentum components appearing in the energy of an atom:

$$\epsilon = \frac{1}{2m}(p_x^2 + p_y^2 + p_z^2). \qquad (6.15)$$

Each quadratic term in the energy contributes $kT/2$ to the internal energy per particle, hence $k/2$ to the specific heat at constant volume. This is known as the *principle of equipartition of energy*.

If our "atoms" are polyatomic molecules, there will be additional quadratic terms in the energy, representing rotational and vibrational degrees of freedom. Each of these terms will contribute $k/2$ to the specific heat additively, as long as the thermal energy $kT$ is sufficient to excite them quantum-mechanically. For example, a diatomic molecule such as $H_2$ has two rotational and one vibrational degrees of freedom. Its energy is of the form

$$\epsilon = \frac{1}{2M}(P_x^2 + P_y^2 + P_z^2) + \frac{J_\perp^2}{2I_\perp} + \frac{J_\parallel^2}{2I_\parallel} + \left(\frac{p^2}{2\mu} + \frac{\mu\omega_0^2}{2}q^2\right), \tag{6.16}$$

where $\mathbf{P}$ is the total momentum, and $J_\perp$ and $J_\parallel$, respectively, denote the angular momentum perpendicular and parallel to the symmetry axis of the molecule. The last two terms represent the energy of a harmonic oscillator corresponding to the vibrational mode. These quantities have to be treated as quantum operators. In particular, the rotational and vibrational energies have discrete eigenvalues, and have minimal energies $kT_{rot}$ and $kT_{vib}$, respectively. Examples of these threshold values are listed in Table 6.1.

At room temperature the thermal energy is not sufficient to excite the vibrational mode of $H_2$, and so its specific heat is $\frac{5}{2}k$. Figure 6.2 shows the heat capacity of 1 mol of molecular hydrogen over a wide range of temperatures on a logarithmic scale. Thus we see quantum mechanics at work even at high temperatures. (See Problem 8.8.)

## 6.4  Distribution of speed

The distribution function is independent of position for a gas in the absence of external potentials. This means that atoms move with the same Maxwellian velocity distribution in every volume element in the air around you. Because of the isotropy of space, the distribution depends only on the magnitude $p$ of the momentum. Thus, the components of the velocity average to zero:

$$\langle \mathbf{v} \rangle = \frac{\int d^3p\, \mathbf{p} f(p)}{m \int d^3p f(p)} = 0. \tag{6.17}$$

Table 6.1 Variational and rotational excitation temperatures (Wilson 1975)

|        | $T_{rot}$ (K) | $T_{vib}$ (K) |
|--------|---------------|---------------|
| $H_2$  | 85.4          | 6100          |
| $N_2$  | 2.86          | 3340          |
| $O_2$  | 2.07          | 2230          |

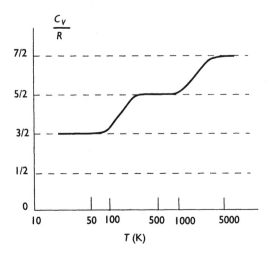

Figure 6.2 Heat capacity per mole of $H_2$ as function of temperature in log scale. After Present (1958).

The mean square velocity is not zero:

$$\langle \mathbf{v}^2 \rangle = \frac{\langle \mathbf{p}^2 \rangle}{m^2} = \frac{\int d^3p\, \mathbf{p}^2 f(\mathbf{p})}{m^2 \int d^3p f(\mathbf{p})} = \frac{3kT}{m} \tag{6.18}$$

and leads to a root-mean-square velocity

$$v_{\mathrm{rms}} = \sqrt{\frac{3kT}{m}}. \tag{6.19}$$

From this we obtain the mean kinetic energy

$$\frac{1}{2}m\langle \mathbf{v}^2 \rangle = \frac{3}{2}kT \tag{6.20}$$

which agrees with the equipartition principle.

The effective volume element in momentum space is $4\pi p^2 \, dp$. The quantity $4\pi p^2 f(p)$ is the distribution of speed, the number of atoms per unit volume per unit interval of $p$ whose magnitude of momentum lies between $p$ and $p + dp$. A qualitative graph of the speed distribution is shown in Fig. 6.3. The area under the entire curve is the particle density $n$. The area under the curve for $p > p_1$ is the density of particles with magnitude of momenta greater than $p_1$. The momentum at the maximum is $p_0 = \sqrt{2mkT}$, and the corresponding velocity is called the *most*

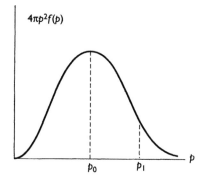

**Figure 6.3** Maxwell–Boltzmann distribution of speed.

*probable velocity:*

$$v_0 = \sqrt{\frac{2kT}{m}}. \tag{6.21}$$

If the gas moves as a whole with uniform velocity $\mathbf{v}_0$, then in maximizing $\Omega\{n_i\}$ we must add the constraint that the average momentum per particle is $\mathbf{p}_0 = m\mathbf{v}_0$. This leads to a momentum distribution centered about $\mathbf{p}_0$:

$$f(\mathbf{p}) = Ce^{-\lambda(\mathbf{p}-\mathbf{p}_0)^2}. \tag{6.22}$$

## 6.5  Entropy

In Chapter 5 we have defined the entropy as

$$S = k \ln \Gamma(E, V) = \sum_{\{n_i\}} \ln \Omega\{n_i\} \tag{6.23}$$

and this yields the correct thermodynamic entropy for an ideal gas. (See Problem 5.7.) Actually, we can just keep the largest term in the sum, and take $S = k \ln \Omega\{\bar{n}_i\}$, for this gives the same answer in the thermodynamic limit. Let us calculate it:

$$\frac{S}{k} = \ln \Omega\{\bar{n}_i\} = N \ln N - \sum_i \bar{n}_i \ln \bar{n}_i$$

$$= N \ln N - V \int d^3p f(\mathbf{p}) \ln f(\mathbf{p}). \tag{6.24}$$

Using the Maxwell–Boltzmann distribution, we have

$$\int d^3 p f(\mathbf{p}) \ln f(\mathbf{p}) = C \int d^3 p \, e^{-\lambda \mathbf{p}^2} \ln \left( C e^{-\lambda \mathbf{p}^2} \right)$$

$$= C \int d^3 p \, e^{-\lambda \mathbf{p}^2} \left( \ln C - \lambda \mathbf{p}^2 \right)$$

$$= (C \ln C) \int d^3 p \, e^{-\lambda \mathbf{p}^2} - \lambda C \int d^3 p \, \mathbf{p}^2 e^{-\lambda \mathbf{p}^2}. \qquad (6.25)$$

Noting that

$$C \int d^3 p \, e^{-\lambda \mathbf{p}^2} = n,$$

$$\lambda C \int d^3 p \, \mathbf{p}^2 e^{-\lambda \mathbf{p}^2} = 4\pi \lambda C \int_0^\infty dp \, p^4 e^{-\lambda p^2}$$

$$= (4\pi \lambda C) \frac{3\sqrt{\pi}}{8\lambda^{5/2}} = \frac{3}{2} n, \qquad (6.26)$$

we obtain

$$\int d^3 p f(\mathbf{p}) \ln f(\mathbf{p}) = n \ln C - \frac{3}{2} n. \qquad (6.27)$$

Putting all this together, we find

$$\frac{S}{Nk} = \ln (V T^{3/2}) + s_0, \qquad (6.28)$$

where $s_0$ is an arbitrary constant. This agrees with that obtained in thermodynamics.

This result shows that, in the thermodynamic limit, almost all states in the microcanonical ensemble have the distribution $\{\bar{n}_i\}$. We shall demonstrate this more directly later.

## 6.6 Derivation of thermodynamics

We can now derive the thermodynamics of an ideal gas. The equation of state $PV = NkT$ was obtained earlier. The internal energy is just the total energy of the system:

$$U(T) = E = \tfrac{3}{2} NkT. \qquad (6.29)$$

Thus $C_V = \tfrac{3}{2} Nk$, and $PV = \tfrac{2}{3} U$. We have calculated the entropy, which can be written in the form

$$\frac{S}{Nk} = \ln V + \frac{3}{2} \ln T. \qquad (6.30)$$

Both $U$ and $S$ are determined only up to an arbitrary additive constant (which may depend on $N$).

Taking the differential of $S$, we have

$$\frac{dS}{Nk} = \frac{dV}{V} + \frac{3}{2}\frac{dU}{U}. \tag{6.31}$$

Using the equation of state to write $dV/V = (P/T)\,dV$, we can rewrite the above as

$$dS = \frac{1}{T}(dU + P\,dV). \tag{6.32}$$

The first law of thermodynamics is a definition of the heat absorbed:

$$dQ = dU + P\,dV. \tag{6.33}$$

The second law is the statement

$$dQ = T\,dS, \tag{6.34}$$

stating that $T$ is the integrating factor that makes $dQ/T$ an exact differential.

That the entropy of an isolated system never decreases is implied by the fact that it is a monotonically increasing function of the volume. For an isolated system, the only thing that can happen is that the volume increases, as when a wall of the container of the gas is suddenly withdrawn.

## 6.7  Fluctuations

We now calculate the mean-square fluctuation about the average occupation $\langle n_i \rangle$, and indicate that it vanishes when $N \to \infty$. This will show that $\langle n_i \rangle$ and $\bar{n}_i$ coincide in that limit.

Starting with

$$\Omega\{n_i\} = \frac{N!}{n_1!n_2!\cdots n_K!}g_1^{n_1}\cdots g_K^{n_K}, \tag{6.35}$$

we now find that the factors $g_i$ are useful for

$$g_k\frac{\partial}{\partial g_k}\Omega\{n_i\} = n_k\Omega\{n_i\}. \tag{6.36}$$

Thus the ensemble average of the occupation number is

$$\begin{aligned}
\langle n_k \rangle &= \frac{g_k\frac{\partial}{\partial g_k}\sum_{\{n_i\}}\Omega\{n_i\}}{\sum_{\{n_i\}}\Omega\{n_i\}}\\
&= g_k\frac{\partial}{\partial g_k}\ln\sum_{\{n_i\}}\Omega\{n_i\}\\
&= \frac{\partial\ln\Gamma}{\partial v_k},
\end{aligned} \tag{6.37}$$

where

$$v_i = \ln g_i. \tag{6.38}$$

The ensemble average of $n_k^2$ is

$$
\begin{aligned}
\langle n_k^2 \rangle &= \frac{\sum_{\{n_i\}} n_k^2 \Omega\{n_i\}}{\sum_{\{n_i\}} \Omega\{n_i\}} = \frac{1}{\Gamma} \frac{\partial^2 \Gamma}{\partial v_k^2} \\
&= \frac{\partial}{\partial v_k} \left( \frac{1}{\Gamma} \frac{\partial \Gamma}{\partial v_k} \right) + \frac{1}{\Gamma^2} \left( \frac{\partial \Gamma}{\partial v_k} \right)^2 \\
&= \frac{\partial}{\partial v_k} \left( \frac{\partial \ln \Gamma}{\partial v_k} \right) + \left( \frac{\partial \ln \Gamma}{\partial v_k} \right)^2 \\
&= \frac{\partial}{\partial v_k} \langle n_k \rangle + \langle n_k \rangle^2.
\end{aligned}
\tag{6.39}
$$

Therefore, the mean-square fluctuation of $n_k$ is

$$\langle n_k^2 \rangle - \langle n_k \rangle^2 = g_k \frac{\partial \langle n_k \rangle}{\partial g_k}. \tag{6.40}$$

The left-hand side can also be written as $\langle [n_k - \langle n_k \rangle]^2 \rangle$.

Assuming for the moment that

$$\langle n_k \rangle \approx \bar{n}_k = g_k C e^{-\beta \epsilon_k}, \tag{6.41}$$

we have

$$g_k \frac{\partial \langle n_k \rangle}{\partial g_k} = g_k C e^{-\beta \epsilon k} = \langle n_k \rangle. \tag{6.42}$$

Thus,

$$\langle n_k^2 \rangle - \langle n_k \rangle^2 = \langle n_k \rangle. \tag{6.43}$$

In the thermodynamic limit $V \to \infty$, the number of cells increase, since they have fixed sizes. The physically relevent quantity is not the occupation of a single cell, but that of a group of cells $\sigma = \sum_i n_i$, such that the fractional occupation $\sigma/N$ approaches a finite limit. We can then show (See prob. 13.3) that

$$\langle (\sigma/N)^2 \rangle - \langle \sigma/N \rangle^2 = N^{-1} \langle \sigma/N \rangle. \tag{6.44}$$

This vanishes like $N^{-1}$ when $N \to \infty$, like a "normal fluctuation".

We conclude that the most probable distribution is almost a certainty, and that the Maxwell–Boltzmann distribution is the most prevalent condition. Suppose all possible states of the gas with given $N$ and $E$ are placed in a jar, symbolically speaking, and you reach in to pick out one state at random. You are almost certain to pick a state which will have the Maxwell–Boltzmann distribution.

The fluctuations cannot be ignored on an atomic length scale, however. In particular, the entropy is subject to fluctuations. The second law of thermodynamics is valid only when these fluctuations can be neglected.

## 6.8  The Boltzmann factor

Our derivation of the most probable distribution does not specifically assume that we are dealing with a classical gas. It only assumes that we have a collection of non-interacting units with energy $\epsilon_i$ for the state $i$. Thus, we have actually proven a more general result:

If a system has possible states labeled by $i$, and the energy of the state $i$ is $\epsilon_i$, then the relative probability for finding the system in state $i$ is given by the Boltzmann factor $e^{-\epsilon_i/kT}$. The absolute probability for the occurrence of the state $i$ is

$$\frac{e^{-\epsilon_i/kT}}{\sum_i e^{-\epsilon_i/kT}}. \tag{6.45}$$

It should be emphasized that $e^{-\epsilon_i/kT}$ is the relative probability for the occurrence of the *state* $i$, not the energy value $\epsilon_i$. The distinction is important, because different states can have the same energy, and this is called a degeneracy in quantum mechanics.

## 6.9  Time's arrow

According to the second law of thermodynamics, the entropy of an isolated system can never decrease. This seems to be a valid conclusion, for most events on the macroscopic scale are irreversible. As we all know, it is useless to cry over spilled milk. Therefore, there appears to be an "arrow of time" that points towards an increase of entropy, and distinguishes past from future. How is this to be reconciled with the time-reversal invariance of the microscopic laws of physics?

Consider Fig. 6.4, which shows two successive frames of a movie of a gas contained in a partitioned box, with a hole through which the particles can pass. Common sense tells us that frame (a) must precede frame (b), and this establishes the arrow of time.

The equations of motion have solutions for which (a) evolves into (b) or vice versa. In fact we can always reverse the history by reversing all the velocities of the particles instantaneously. However, the situations (a) and (b) are not symmetrical. As an initial state, almost any state that looks like (a) will evolve into a uniform state like (b). But an initial state (b) will not evolve into (a), unless it is very carefully prepared to achieve that purpose.

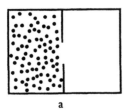 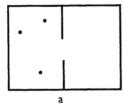 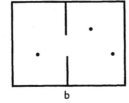

a        b

*Figure 6.4* If these are two successive frames of a movie, common sense tells us that (a) must precede (b), and this establishes "time's arrow."

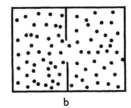

a        b

*Figure 6.5* In this case, one cannot tell which frame precedes which.

The point is that, in the $3N$-dimensional phase space of the system, the initial conditions that will make (b) evolve into (a) have a negligible measure, when $N$ is large. On the other hand, most non-uniform states like (a) will develop into a uniform state like (b), because such evolutions are overwhelmingly favored by phase space.

If we start with an "average" initial condition, (b) has a chance to evolve into (a) only after the order of $e^N$ collision times. (See Problem 5.11.) For $N \approx 100$ as depicted in Fig. 6.4, this time is much longer than the age of the universe, which is a mere $10^{17}$ s.

If $N$ is small, as illustrated in Fig. 6.5, then the system can evolve from (a) to (b) or vice versa with equal probability, and there is complete reversibility. Time's arrow, therefore, is not inherent in the equation of motion, but is the property of a large-$N$ system set by the initial condition. The spilling of milk is irreversible, because someone had prepared the milk in the extremely improbable state of being inside the milk bottle.

In the case of the whole universe, unless the Big Bang was "prepared" in any special way, we must conclude that time's arrow signifies the spontaneous breaking of time-reversal invariance. On the other hand, it is believed that the interaction of elementary particles contains an extremely weak component that violates time-reversal invariance. It is possible that this could have tilted the Big Bang along a preferred time axis.

## Problems

**6.1**  The energy of individual atoms in a gas fluctuates about an average value $\frac{3}{2}kT$ because of collisions.

(a)  Verify this by calculating $\bar{\epsilon}$, the mean of $\epsilon = \mathbf{p}^2/2m$, with respect to the Maxwell–Boltzmann distribution.

(b)  Show that $\overline{\epsilon^2} - \bar{\epsilon}^2 = \frac{3}{2}(kT)^2$.

**6.2**  Find the energy distribution function $P(E)$ for a classical non-relativistic ideal gas, such that $P(E)\,dE$ is the density of atoms with energy between $E$ and $E + dE$.

**6.3**  If there is an external potential $U(\mathbf{r})$, then the distribution function is non-uniform in space:

$$f(\mathbf{p}, \mathbf{r}) = Ce^{-[(\mathbf{p}^2/2m)+U(\mathbf{r})]/kT}.$$

A column of gas under gravity has a temperature independent of height $z$. Show that the density as a function of height is given by

$$n(z) = n(0)e^{-mgz/kT}$$

where $g$ is the acceleration due to gravity.

**6.4**  In the atmosphere the temperature varies with height. Assume that there is a steady-state adiabatic convection, i.e. no heat transfer in the vertical direction.

(a)  Show that the temperature $T(z)$ changes with height $z$ according to

$$k\frac{dT}{dz} = -\frac{\gamma - 1}{\gamma}mg.$$

Find $T(z)$ and the altitude of the top of the atmosphere where the temperature becomes zero.

(b)  Show that the pressure $P(z)$ changes with height according to

$$\frac{dP}{P} = -\frac{mg}{kT(z)}\,dz.$$

Integrate this to find $P(z)$.

**6.5**  A gas in equilibrium has a distribution function

$$f(\mathbf{p}, \mathbf{r}) = C_0(1 + \gamma x)\,(2\pi mkT)^{-3/2}\exp\left(-\mathbf{p}^2/2mkT\right),$$

where $x$ is the distance along an axis with a fixed origin, $\gamma$ is a constant. What is the nature of the force acting on the gas?

**6.6**  Molecules in a centrifuge rotate about an axis at constant angular velocity $\omega$. They are at rest in the rotating frame, but experience a centrifugal force $m\omega^2 r$, where $r$ is the normal distance from the axis. This is equivalent to an external potential

$$U(r) = -\tfrac{1}{2}m\omega^2 r^2.$$

Two dilute gases of molecular masses $m_1$ and $m_2$ respectively, are placed in a centrifuge rotating at a circular frequency $\omega$. Derive the ratio $n_1/n_2$ of their densities as a function of the distance $r$ from the axis of rotation.

**6.7** The Maxwell–Boltzmann distribution for a relativistic gas is

$$f(\mathbf{p}) = Ce^{-\sqrt{p^2+m^2}/kT},$$

where we use units in which the velocity of light is $c = 1$.

(a) Find the most probable velocity. Obtain its non-relativistic ($kT \ll m$) and ultra-relativistic ($kT \gg m$) limits, both with first-order corrections.
(b) Set up an expression for the pressure. Show that $PV = U/3$ in the ultra-relativistic limit, where $U$ is the average energy.
(c) Find the velocity distribution function $f(\mathbf{v})$, such that $f(\mathbf{v})\,d^3v$ is the density of particles whose velocity lies in the volume element $d^3v$. Find the non-relativistic limit to first order in $v/c$.
(d) At what temperatures would relativistic effects be important for a gas of $H_2$ molecules?

**6.8** The Doppler formula for the observed frequency $f$ from a source moving with velocity $v_x$ along the line of sight of an observer is

$$f = f_0\left(1 + \frac{v_x}{c}\right),$$

where $f_0$ is the frequency in the rest frame of the source.

(a) What is the distribution in frequency of a particular spectral line radiated from a gas at temperature $T$?
(b) Find the breadth of the line, defined as the variance $\overline{(f - f_0)^2}$.
(c) Atomic hydrogen and atomic oxygen are both present in a hot gas. How much broader is the hydrogen line compared to the oxygen line, of roughly the same frequency?

**6.9** Neutrinos are particles whose energy–momentum relation is given by $\epsilon(p) = cp$, where $c$ is the velocity of light. Consider $N$ neutrinos in a volume $V$, at a temperature $T$ sufficiently high that the system can be treated as a classical gas.

(a) Find the heat capacity $C_V$ of the system.
(b) Find the pressure of the system in terms of the internal energy $U$. Give it in terms of the temperature.

**6.10** If we integrate the Maxwell–Boltzmann distribution over from some momentum up, we are faced with error functions

$$erfc(y) = \frac{2}{\sqrt{\pi}} \int_y^\infty dx.$$

[See *Handbook of Mathematical Functions*, M. Abramowitz and I.A. Stegun, Eds. (National Bureau of Standards, 1964).]

(a)  Show the following asymptotic behavior for large $y$:

$$\int_y^\infty dx\, e^{-x^2} \approx e^{-y^2} \left[ \frac{1}{2y} - \frac{1}{4y^3} + \frac{3}{8}\frac{1}{y^5} + \cdots \right].$$

(*Hint*: Transform to new integration variable $t = x^2$. The asymptotic behavior of the integral $\int_{y^2}^\infty dt\, t^{-1/2}\, e^{-t}$ can be obtained by repeated partial integrations.)

(b)  Differentiate $\int_y^\infty dx\, e^{-\lambda x^2}$ with respect to $\lambda$, and then set $\lambda = 1$, to obtain the asymptotic formulas

$$\int_y^\infty dx\, x^2 e^{-x^2} \approx \frac{e^{-y^2}}{2y} (y^2 + 1/2),$$

$$\int_y^\infty dx\, x^4 e^{-x^2} \approx \frac{e^{-y^2}}{2y} (y^4 + y^2 + 3/4).$$

**6.11**  Suppose a surface of the container of a gas absorbs all molecules striking it with a normal velocity greater than $v_0$. Find the absorption rate $W$ per unit area.

**6.12**  The atmosphere contains molecules with high velocities that can escape the earth's gravitational field.

(a)  What fraction of the $H_2$ gas at sea level, at temperature 300 K, can escape from the earth's gravitational field?

(b)  Give an order-of-magnitude estimate of the time needed for the escape, on the basis of a random walk.

**6.13**  A gas of $N$ atoms was initially in equilibrium in a volume $V$ at temperature $T$. In an evaporation process, all atoms with energy greater than $\epsilon_0 = p_0^2/2m$ were allowed to escape, and the gas eventually reestablishes a new equilibrium. Assume $y \equiv \epsilon_0/kT \gg 1$.

(a)  Find the change $\Delta N$ in the number of atoms, and the change $\Delta E$ in the energy of the gas, as functions of $\epsilon_0$.

(b)  Find the fractional change in temperature $\Delta T/T$ as a function of $\Delta N/N$. (*Hint*: Find $\Delta T$ via $E/N = \frac{3}{2}kT$. Express $\epsilon_0$ in terms of $\Delta N$ using an iterative process assuming the smallness of the latter.)

**6.14**  The following exercises illustrate the equipartition of energy.

(a)  A long thin needle floats in a gas at constant temperature. On the average, is its angular momentum vector nearly parallel to or perpendicular to the long axis of the grain? Explain.

(b)  A capacitor, $C = 100\,\mu\text{F}$, in a passive circuit (no driving voltage) is at temperature $T = 300$ K. Calculate the rms voltage fluctuation.

**6.15** Consider an insulated space ship which is a cylinder of length $L$ and cross section $A$ filled with air (treated as $N_2$ gas) at STP. It is brought to a sudden stop from an initial velocity of 7 km/s.

(a) Assuming that a good fraction of the original translational kinetic energy of the air was converted to heat, estimate the temperature rise inside the space ship. Will the astronauts be fried?

(b) Suppose the space ship was stopped by constant deceleration $a$ parallel to the axis of the cylinder, such that the air was in local equilibrium. Show that the pressure difference between the front and back of the space ship is given by

$$\Delta P = P_0 \left(1 - e^{-maL/kT}\right),$$

where $m$ is the mass of an air molecule.

(c) Assuming $maL/kT \ll 1$, find the force $F$ exerted on the space ship by the air inside, the total work $W$ done by the air, and the temperature rise inside the space ship.

(d) What condition must be imposed on the deceleration to allow the establishment of local equilibrium?

# Chapter 7

# Transport phenomena

## 7.1 Collisionless and hydrodynamic regimes

A gas tends towards thermal equilibrium through atomic collisions, which can transport mass, momentum, and energy from one part of the system to another. On average, these quantities are transported over a mean free path $\lambda$ per collision. An important parameter in the description of transport phenomena is the ratio $\lambda/L$, where $L$ is an external length scale, such as the size of the container, or the wavelength of a density variation. Two extreme limits are amenable to analytical treatment:

Collisionless regime:     $\lambda \gg L$,
Hydrodynamic regime:   $\lambda \ll L$.

We can illustrate the two cases by looking at how a gas flows through a hole in the wall, of dimension $L$, as shown in Fig. 7.1.

When $\lambda \gg L$, the atom that went through the hole came from a last collision very far from the hole, and will not collide with another atom again until it gets very far from the hole. The passage through the hole can be described by ignoring collisions. A practical example is air leaking into a vacuum system through a very small crack.

When $\lambda \ll L$, an atom makes many collisions during the passage, and thermalizes with the local atoms during its journey. As a consequence, it moves as part of a collective flow. This regime is described by hydrodynamics.

In the collisionless regime, atoms escape through a hole in the wall through *effusion*. Let us set up the $x$-axis normal to the area of the hole. The flux of atoms through the hole, due to those with momenta lying in the element $d^3p$, is given by

$$dI = v_x f(\mathbf{p})\, d^3p. \tag{7.1}$$

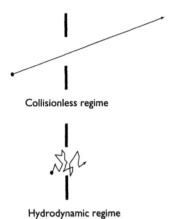

Collisionless regime

Hydrodynamic regime

*Figure 7.1* Different regimes corresponding to different magnitudes of the mean free path relative to physical dimensions.

The total flux is then

$$I = \int_{v_x>0} v_x f(\mathbf{p}) \, d^3p$$

$$= C \int_0^\infty dp_x \frac{p_x}{m} e^{-\lambda p_x^2} \left[ \int_{-\infty}^\infty dp_y \, e^{-\lambda p_y^2} \right]^2 , \qquad (7.2)$$

where

$$\lambda = (2mkT)^{-1},$$

$$C = n \, (2\pi mkT)^{-3/2} . \qquad (7.3)$$

We easily obtain the result

$$I = n \sqrt{\frac{kT}{2\pi m}}. \qquad (7.4)$$

Since the flux is proportional to $v_x$, fast atoms have a higher flux than slow ones, and so the escaped gas has a higher average energy per particle than the gas left behind. If the original volume contains a mixture of two gases at the same temperature, with atomic masses $m_1, m_2$, and densities $n_1, n_2$, then they effuse out of a small hole at different rates, with a ratio

$$\frac{I_1}{I_2} = \frac{n_1}{n_2} \sqrt{\frac{m_2}{m_1}}. \qquad (7.5)$$

This formula is the basis for a method to separate nuclear isotopes.

In the hydrodynamic regime, where $\lambda \ll L$, the gas reaches local thermal equilibrium over a distance small compared to $L$, but large compared to $\lambda$. It has well-defined local properties that vary slowly in space and in time:

$T(\mathbf{r}, t)$: Local temperature

$n(\mathbf{r}, t)$: Local density  (7.6)

$\mathbf{u}(\mathbf{r}, t)$: Local flow velocity

The flow trajectory with $\mathbf{u}$ as tangent vector is called a "streamline". In general the system settles into local thermodynamic equilibrium rather quickly, in the order of a collision time, but it takes much longer for the system to approach a uniform state. The local equilibrium is described by a local Maxwell–Boltzmann distribution

$$f(\mathbf{p}, \mathbf{r}, t) = Cn(\mathbf{r}, t) \exp\left[ -\frac{|\mathbf{p} - m\mathbf{u}(\mathbf{r}, t)|^2}{2mkT(\mathbf{r}, t)} \right], \qquad (7.7)$$

where the local variables are governed by the equations of hydrodynamics.

## 7.2 Maxwell's demon

As we have noted, effusion is a velocity filter, but it acts in the same manner in both directions. A velocity filter that acts only in one direction would violate the second law of thermodynamics.

Maxwell imagined a way to do this by postulating a "demon" who operates a trap door in a wall separating two gases $A$ and $B$, which were initially in equilibrium at the same temperature. The demon opens the door and allows fast atoms to go from $A$ to $B$, but not the slow ones, and allows slow atoms to go from $B$ to $A$, but not the fast ones. As time goes on, the average energy in $A$ will rise, while that in $B$ will fall. Thus the temperature of $A$ will rise and that of $B$ will fall "spontaneously". This fanciful idea has provoked much debate, centering on whether the demon should be considered part of the system, whether he/she/it has entropy, etc. Szilard (1929) pointed out that the demon needs information concerning the velocity of the approaching atoms, and that the second law can be preserved by regarding information as negative entropy. This idea has blossomed into the field of information theory, with applications to the theory of computation (Leff 1990).

## 7.3 Non-viscous hydrodynamics

The hydrodynamic regime is based on the smallness of $\lambda/L$, where $L$ refers to the characteristic wavelength of spatial variations. To the lowest order, we consider the limit $\lambda/L \to 0$, in which collisions are neglected, and the changes in the local variables are governed solely by conservation laws. This leads to non-viscous hydrodynamics. In this limit the local equilibrium persists indefinitely; there is no damping mechanism for it to decay to a uniform state.

The relevant conservation laws are those for number of particles, and for momentum and energy. Because of particle conservation, the mass density

$$\rho(\mathbf{r}, t) = mn(\mathbf{r}, t) \tag{7.8}$$

satisfies the equation of continuity

$$\frac{\partial \rho}{\partial t} + \nabla \cdot (\rho \mathbf{u}) = 0 \quad \text{(Continuity equation)}. \tag{7.9}$$

The conservation of momentum is expressed through Newton's equation $F = ma$ in a local frame co-moving with the gas along a streamline. Consider an element of the gas contained between $x$ and $x + dx$, in a small cylinder of normal cross section $A$. In the absence of collisions, the $x$-component of the force acting on it due to neighboring gas elements arises purely from hydrostatic pressure:

$$dF_x = [P(x) - P(x + dx)] A = -A \frac{\partial P}{\partial x} dx, \tag{7.10}$$

where $P(x)$ is the local pressure. This is the total net force on the element if we neglect collisions, which can create a shear force that leads to viscosity.

Newton's equation now states

$$-A \frac{\partial P}{\partial x} dx = \frac{du_x}{dt} dm, \tag{7.11}$$

where the mass element $dm$ is given by

$$dm = A\rho \, dx. \tag{7.12}$$

Thus, we have

$$\rho \frac{du_x}{dt} + \frac{\partial P}{\partial x} = 0, \tag{7.13}$$

where $du_x/dt$ is evaluated in the co-moving frame. In a fixed frame in the laboratory, it is given by

$$\frac{du_x}{dt} = \frac{\partial u_x}{\partial t} + \left( u_x \frac{\partial}{\partial x} + u_y \frac{\partial}{\partial y} + u_z \frac{\partial}{\partial z} \right) u_x. \tag{7.14}$$

Generalizing the above considerations to any component of $\mathbf{u}$, and adding an external force per unit volume $\mathbf{f}^{\text{ext}}$, we obtain *Euler's equation*

$$\rho \left( \frac{\partial}{\partial t} + \mathbf{u} \cdot \nabla \right) \mathbf{u} + \nabla P = \mathbf{f}^{\text{ext}} \quad \text{(Euler's equation)}. \tag{7.15}$$

When collisions are ignored, there is no mechanism for energy transfer between a gas element and its neighbors. This means a gas element can only undergo adiabatic transformations in a co-moving frame along a streamline. For an ideal gas we have

$$\left(\frac{\partial}{\partial t} + \mathbf{u} \cdot \nabla\right)\left(P\rho^{-\gamma}\right) = 0 \quad \text{(Adiabatic condition)}, \tag{7.16}$$

where $\gamma = C_P/C_V$, and the local equation of state gives $P = \rho kT/m$.

To be consistent with the premise $\lambda/L \to 0$, we must assume small deviations from equilibrium. Accordingly, we retain the local velocity $\mathbf{u}$, and all spatial and time derivatives, only to first order. Second order quantities, defined as products of first-order quantities, will be neglected. In particular we put

$$\nabla \cdot (\rho \mathbf{u}) = \rho \nabla \cdot \mathbf{u} + \mathbf{u} \cdot \nabla \rho \approx \rho \nabla \cdot \mathbf{u}, \tag{7.17}$$

because $\mathbf{u} \cdot \nabla \rho$ is of second-order smallness. This approximation leads to the linearized equations of non-viscous hydrodynamics:

$$\frac{\partial \rho}{\partial t} + \rho \nabla \cdot \mathbf{u} = 0,$$

$$\rho \frac{\partial \mathbf{u}}{\partial t} + \nabla P = \mathbf{f}^{\text{ext}}, \tag{7.18}$$

$$\frac{\partial}{\partial t}\left(P\rho^{-\gamma}\right) = 0,$$

where $P = \rho kT/m$.

## 7.4  Sound waves

Differentiating the first hydrodynamics equation with respect to time and retaining only first-order terms, we obtain

$$\frac{\partial^2 \rho}{\partial t^2} + \rho \nabla \cdot \frac{\partial \mathbf{u}}{\partial t} = 0, \tag{7.19}$$

where a term $(\partial \rho/\partial t)\nabla \cdot \mathbf{u}$ has been neglected, because it is the product of two derivatives, and thus of second order. Substituting $\partial \mathbf{u}/\partial t$ from the second hydrodynamics equation with $\mathbf{f}_{\text{ext}} = 0$, we have

$$\frac{\partial^2 \rho}{\partial t^2} - \rho \nabla \cdot \left(\frac{1}{\rho}\nabla\right) P = 0. \tag{7.20}$$

To first order, this is equivalent to

$$\frac{\partial^2 \rho}{\partial t^2} - \nabla^2 P = 0. \tag{7.21}$$

To evaluate $\nabla^2 P$, we make use of the third hydrodynamics equation:

$$\nabla^2 P = \nabla \cdot \nabla P = \nabla \cdot \left[ \left( \frac{\partial P}{\partial \rho} \right)_S \nabla \rho \right]$$

$$\approx \left( \frac{\partial P}{\partial \rho} \right)_S \nabla^2 \rho = \rho \kappa_S \nabla^2 \rho, \tag{7.22}$$

where $\kappa_S$ is the adiabatic compressibility. Thus we have

$$\nabla^2 \rho - \frac{1}{c^2} \frac{\partial^2 \rho}{\partial t^2} = 0, \tag{7.23}$$

where

$$c = \frac{1}{\sqrt{\rho \kappa_S}}. \tag{7.24}$$

This gives a wave equation for a sound wave of velocity $c$. Taking collisions into account will lead to the damping of sound.

## 7.5   Diffusion

In the next approximation, we take collisions into account to first order in $\lambda/L$, and this leads to transport phenomena. We shall not go into systematic treatments, but only give heuristic discussions, beginning with diffusion.

Suppose the density of a gas at points 1 and 2 along the $x$-axis are $n_1$ and $n_2$, respectively. Atomic collisions tend to iron out the density variation, because there is a higher flux of particles going from the high density to the low density region, as indicated in Fig. 7.2. This gives rise to *diffusion*. This is sometimes called "self-diffusion", to distinguish it from "mutual diffusion", where one atomic species diffuses in the medium of another species. We assume that the temperature is uniform, so that the atoms have a most probable velocity $\bar{v} = \sqrt{2kT/mT}$. On average, 1/6 of the atoms travel along the positive $x$-axis, and 1/6 along the negative

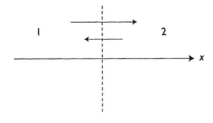

Figure 7.2 Diffusion results from the flux of particles being more in one direction than the other, due to a density gradient.

$x$-axis, in 3D space. The flux of particles from 1 to 2 is therefore $n_1\bar{v}/6$, that from 2 to 1 is $n_2\bar{v}/6$, and the net flux from 1 to 2 is $(n_1 - n_2)\bar{v}/6$.

The densities at 1 and 2 can affect each other only if the separation is larger than a mean free path. Assume that they are separated by a distance $r_0$ of the order of the mean free path. The $x$-component of the particle current density can be written as

$$j_x \approx -\frac{r_0\bar{v}}{6}\frac{\partial n}{\partial x}. \tag{7.25}$$

The minus sign occurs because a positive gradient along the $x$-axis drives particles in the negative direction. In vector notation, we have

$$\mathbf{j} = -D\nabla n,$$

$$D = \frac{r_0\bar{v}}{6}, \tag{7.26}$$

where $D$ is a transport coefficient called the *diffusion constant*.

The particle current satisfies the continuity equation

$$\frac{\partial n}{\partial t} + \nabla \cdot \mathbf{j} = 0. \tag{7.27}$$

Using $\mathbf{j}$ obtained earlier, we obtain the *diffusion equation*:

$$\frac{\partial n(\mathbf{r},t)}{\partial t} - D\nabla^2 n(\mathbf{r},t) = 0. \tag{7.28}$$

The solution, with the initial condition that there were $N$ particles at $\mathbf{r} = 0$, is

$$n(\mathbf{r},t) = \frac{N}{(4\pi D)^{3/2}}\frac{e^{-r^2/(4\pi Dt)}}{t^{3/2}}, \tag{7.29}$$

where $r = |\mathbf{r}|$. The initial condition is

$$n(\mathbf{r},t) \xrightarrow[t\to 0]{} N\delta^3(\mathbf{r}). \tag{7.30}$$

Conservation of particles is expressed through the fact

$$\int d^3r\, n(\mathbf{r},t) = N. \tag{7.31}$$

As time goes on, the particles diffuse out from the origin, forming a Gaussian distribution with an expanding width $\sqrt{4\pi Dt}$.

Using $\bar{v} = \sqrt{2kT/m}$, $r_0 = 2\lambda = 2/(n\sigma)$, where $\sigma$ is the collision cross section, we obtain

$$D \approx \frac{1}{3n\sigma}\sqrt{\frac{2kT}{m}}, \tag{7.32}$$

which is an order-of-magnitude estimate.

The phenomenon of diffusion can be approached from different perspectives. It can be treated in terms of random walk (Problem 5.4), Brownian motion (Section 16.4), and general stochastic processes (Section 17.5, Problem 18.6).

## 7.6  Heat conduction

Assume now that the density $n$ is uniform, while the temperature $T$ varies slowly in space. The flux of particles in Fig. 7.2 is now equal to $n\bar{v}/6$ in both directions. The average kinetic energy per particle is $lkT/2$, where $l$ is the number of degrees of freedom. The heat flux from 1 to 2 is given by the flux of thermal energy

$$q_x = \frac{n\bar{v}l}{12}(kT_1 - kT_2) \approx -\frac{n\bar{v}r_0 lk}{12}\frac{\partial T}{\partial x}. \tag{7.33}$$

We can write, in vector notation,

$$\mathbf{q} = -\kappa \nabla T,$$

$$\kappa = \frac{1}{6}n\bar{v}r_0 c_V, \tag{7.34}$$

where $c_V = lk/2$, and $\kappa$ is a transport coefficient called the *coefficient of thermal conductivity*. Taking $\bar{v} = \sqrt{2kT/m}$, $r_0 = 2\lambda = 2/(n\sigma)$, we have the order-of-magnitude estimate

$$\kappa = \frac{c_V}{3n\sigma}\sqrt{\frac{2kT}{m}}. \tag{7.35}$$

When no work is performed, the heat absorbed by an element is equal to the increase in its internal energy, according to the first law of thermodynamics. This leads to the conservation law

$$\nabla \cdot \mathbf{q} + \frac{\partial u}{\partial t} = 0, \tag{7.36}$$

where $u$ is the internal energy per unit volume. Using (7.34), we have

$$\frac{\partial u}{\partial t} - \kappa \nabla^2 T = 0. \tag{7.37}$$

For an ideal gas $u = nc_V T$, where $c_V$ is the specific heat per particle. Thus we have a diffusion equation called the *heat conduction equation*:

$$\frac{\partial T}{\partial t} - \frac{\kappa}{nc_V}\nabla^2 T = 0. \tag{7.38}$$

## 7.7 Viscosity

The velocity of a gas flowing past a wall has a profile as illustrated in Fig. 7.3. Here, the gas is flowing along the $x$-direction with a non-uniform flow velocity $u_x(y)$. The gas sticks to the wall at $y = 0$, as expressed by the fact $u_x(0) = 0$. We assume that the density and temperature are uniform.

Consider a plane normal to the $y$-axis, shown as the dotted line in Fig. 7.3. The gas above this plane experiences a frictional force per unit area $F_y$ given empirically by

$$F_y = -\nu \frac{\partial u_x(y)}{\partial y} \tag{7.39}$$

which defines $\nu$, the *coefficient of viscosity*. This force is caused by a transport of "$x$-component momentum" along the $y$-direction, from 1 to 2 in Fig. 7.3. We now calculate $\nu$, using the theoretical definition

$$F_y = \text{Flux of "$x$-component momentum" along the $y$-direction.} \tag{7.40}$$

Note that $u_x$ is the average collective flow velocity, and the transported momentum is that of the collective flow. The individual molecules, of course, dart about in all directions with average speed $\bar{v}$ relative to the flow velocity. Since the "$x$-component momentum" per particle is $mu_x$, and the flux of particles in the $y$-direction is $n\bar{v}/6$, we have

$$F_y = \frac{1}{6} n\bar{v}m[u_x(y_1) - u_x(y_2)]. \tag{7.41}$$

Choosing the points 1 and 2 to be separated by a distance $r_0$ of the order of a mean free path, we obtain

$$\nu = \frac{1}{6} n\bar{v}mr_0. \tag{7.42}$$

Putting $r_0 = 2\lambda$, we have the estimate

$$\nu = \frac{\sqrt{2mkT}}{3\sigma}. \tag{7.43}$$

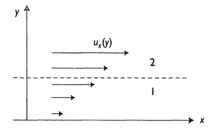

*Figure 7.3* Particles sticking to the wall cause viscosity.

Note that this is independent of the density, a surprising prediction borne out by experiments on gases.

## 7.8  Navier–Stokes equation

When viscosity is taken into account, the forces acting on a fluid element are no longer normal to the surfaces of the element. The hydrostatic pressure $P$ is now generalized to a pressure tensor $P_{ij}$, which gives the $j$th component of the force per unit area acting on the $i$th face, as illustrated in Fig. 7.4 The vector force $\mathbf{F}_i$ on the $i$th face has the components indicated below:

$$
\begin{aligned}
\mathbf{F}_1 &= A(P_{11}, P_{12}, P_{13}), \\
\mathbf{F}_2 &= A(P_{21}, P_{22}, P_{23}), \\
\mathbf{F}_3 &= A(P_{31}, P_{32}, P_{33}),
\end{aligned}
\tag{7.44}
$$

where $A$ is the surface area. We can thus split the pressure tensor into two terms:

$$
P_{ij} = \delta_{ij} P + P'_{ij},
\tag{7.45}
$$

where the off-diagonal term $P'_{ij}$ depends on the viscosity, and turns out to have the form (Huang 1987, Section 5.8)

$$
P'_{ij} = -v \left[ \frac{\partial u_i}{\partial x_j} + \frac{\partial u_j}{\partial x_i} - \frac{2}{3}\delta_{ij} \nabla \cdot \mathbf{u} \right].
\tag{7.46}
$$

The generalization of Euler's equation reads

$$
\rho \left( \frac{\partial}{\partial t} + \mathbf{u} \cdot \nabla \right) u_i + \frac{\partial P_{ij}}{\partial x_j} = f_i^{\text{ext}},
\tag{7.47}
$$

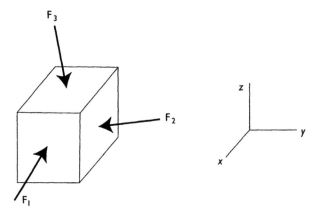

*Figure 7.4* Due to viscosity, forces acting on a liquid element are not normal to the surface of the element, but have a shear component.

or, more explicitly,

$$\rho \left( \frac{\partial}{\partial t} + \mathbf{u} \cdot \nabla \right) \mathbf{u} + \nabla \left( P - \frac{\nu}{3} \nabla \cdot \mathbf{u} \right) - \nu \nabla^2 \mathbf{u} = \mathbf{f}^{\text{ext}}, \tag{7.48}$$

where $\mathbf{f}^{\text{ext}}$ is the external force per unit volume. This is known as the *Navier–Stokes equation*.

The existence of viscosity furnishes a new scale in hydrodynamics. The dimensionality of viscosity is

$$\nu \sim \frac{\text{force/area}}{\text{velocity/length}} \sim \frac{mu}{tL^2} \frac{L}{u} \sim \frac{m}{Lt}, \tag{7.49}$$

where $m = $ mass, $u = $ velocity, $L = $ length, $t = $ time. Using $\nu$ we can define a dimensionless quantity characterizing hydrodynamic flow called the *Reynolds number*:

$$R = \frac{\rho L u_0}{\nu}, \tag{7.50}$$

where

$$\begin{aligned} \rho &= \text{mass density,} \\ L &= \text{characteristic length,} \\ u_0 &= \text{flow velocity,} \\ \nu &= \text{viscosity.} \end{aligned} \tag{7.51}$$

For a stationary object of size $L$, immersed in a fluid of mass density $\rho$, and viscosity $\nu$, flowing with velocity $u_0$, this number marks the onset of turbulence. When $R \ll 1$ the fluid flows past the object in streamline flow, and when $R \gg 1$ we have turbulent flow.

## Problems

**7.1** A high-vacuum chamber develops a small crack of area $\sigma$, and air from the outside leaks in by effusion.

(a) Find the rate of air molecules leaking in through the crack.

(b) After a very short time, the leak was discovered, and patched up by diligent students. The small amount of gas came to equilibrium inside the chamber. Show that its absolute temperature is higher than that of the air outside by a factor 4/3. (*Hint*: Suppose the leak existed for a time $\tau$. Calculate the energy $E$ and number of molecules $N$ that got through during that time. The equilibrium temperature inside the chamber is determined by the ratio $E/N$.)

**7.2** Natural uranium ore contains isotopes $^{238}U$ and $^{235}U$ with abundances 99.27% and 0.73%, respectively. To increase the relative abundance of $^{235}U$, a sample of natural uranium is vaporized, and made to effuse successively into a series of vacuum chambers. How many stages are required to achieve equal abundance in the two isotopes?

**7.3**   Show that the velocity of sound in an ideal gas is

$$c = \sqrt{\frac{kT\gamma}{m}},$$

where $\gamma = C_P/C_V$. Evaluate this for air (nitrogen) at STP.

**7.4**   Sound propagates adiabatically because the effect of heat conduction can be neglected. Verify this in a real gas, as follows.

Consider a sound wave of wavelength $L$ and period $\tau = L/c$, where $c$ is the sound velocity. Let $\Delta T$ be the variation in temperature over $L$. Then the magnitude of the heat flux is $q \approx K\Delta T/L$, where $K$ is the coefficient of thermal conductivity. The amount of heat transferred by conduction across unit area over the distance $L$ is, therefore, $Q_1 = q\tau = K\Delta T/c$. The amount of heat needed to equalize the temperature is $Q_2 = C_P\Delta T$. Thus, heat conductivity may be ignored if $Q_1 \ll Q_2$.

Test this condition for a sound wave with $L = 10$ ft in air at STP, with the following data:

$$\rho = 0.08 \text{ lb ft}^{-3},$$

$$c = 1088 \text{ ft s}^{-1},$$

$$C_P = 0.24 \text{ Btu lb}^{-1} \, (^\circ\text{F})^{-1},$$

$$K = 0.0157 \text{ Btu hr}^{-1} \text{ ft}^{-1} \, (^\circ\text{F})^{-1}.$$

**7.5**   Rederive the equation for a sound wave, using the Navier–Stokes equation instead of the Euler equation. Find the damping coefficient for sound.

**7.6**   A long thin tube along the $x$-axis contains a gas of $N$ particles initially concentrated at $x = 0$. A particle detector is placed at $x = L$. At what time does the detector register the first signal?

**7.7**   A gas is sealed between the thermal panes of a window. The insulating power of the window is taken to be the inverse of the coefficient of thermal conductivity of the gas. Normally the gas is air, of average molecular weight 30. What should be the molecular weight of the gas, in order to double the insulating power? *Hint*: Consider how the coefficient of thermal conductivity depends on molecular weight, through the mass of the molecule and the collision cross section.

**7.8**   Heat is generated uniformly in the Earth's interior due to radioactivity, at the rate $W$ cal g$^{-1}$s$^{-1}$. Assume that the Earth is spherically symmetric with radius $R$, uniform mass density $\rho$, and coefficient of thermal conductivity $\kappa$. Ignore the effect of all other heat sources.

Find the temperature $T(r)$, as a function of distance $r$ from the center of Earth.

**7.9**   During heat transfer characterized by the heat flux vector **q**, there is both entropy flow and irreversible entropy production.

(a)   Show that the entropy density $s$ satisfies the equation

$$\frac{\partial s}{\partial t} + \frac{1}{T}\nabla \cdot \mathbf{q} = 0,$$

where $T$ is the absolute temperature.

(b)   Show that the rate of entropy production is given by $R_s = \mathbf{q} \cdot \nabla (1/T)$, hence

$$R_s = \kappa \left(\frac{\nabla T}{T}\right)^2.$$

**7.10**   The temperature of a pond is just above freezing. The temperature of the air suddenly drops by $\Delta T$, and a sheet of ice begins to form at the surface of the pond, and thickens as time goes on. Find the rate at which the thickness of the ice sheet grows. (Consider only the transition of water to ice and ignore the cooling of the ice once it is formed. Let the latent heat of ice be $\ell$, the mass density of water $\rho$, and the coefficient of thermal conductivity of ice $\kappa$.)

# Chapter 8

# Quantum statistics

## 8.1  Thermal wavelength

Atoms in a gas are actually wave packets. They can be pictured as billiard balls at high temperatures, because their size is much smaller than the average interparticle distance. As the temperature decreases, however, the wave packets begin to spread, and when they begin to overlap with each other, specific quantum effects must be taken into account.

The spatial extension of a wave packet is governed by the de Broglie wavelength $\lambda_0 = h/p_0$, where $p_0$ is the average momentum. For a gas in equilibrium at temperature $T$, it is given through

$$\frac{p_0^2}{2m} = \frac{3}{2}kT, \tag{8.1}$$

which gives $\lambda_0 = h/\sqrt{3mkT}$. We can make a wave packet by superimposing plane waves whose wavelengths lie in the neighborhood of $\lambda_0$. The spatial extension $\Delta x$ and the momentum spread $\Delta p$ must satisfy the uncertainty relation $\Delta x \Delta p \sim \hbar$. There is freedom to adjust $\Delta x$ and $\Delta p$, but for reasonable choices $\lambda_0$ serves as an order-of-magnitude estimate of the degree of spatial localization.

We define the *thermal wavelength* by

$$\lambda = \sqrt{\frac{2\pi\hbar^2}{mkT}}, \tag{8.2}$$

where the numerical factors are chosen to give a neater appearance to some formulas. For a system to be in the classical regime, $\lambda$ should be much smaller than the average interparticle distance $r_0$. Since the latter is proportional to $n^{1/3}$, where $n$ is the density, the condition can be stated as

$$n\lambda^3 \ll 1 \quad \text{(Classical regime)}. \tag{8.3}$$

Quantum effects become important at temperatures lower than the "degeneracy temperature" $T_0$ corresponding to

$$n\lambda^3 \approx 1 \quad \text{(Onset of quantum effects)}. \tag{8.4}$$

At this point, the wave functions of different atoms begin to overlap, and we must treat the system according to quantum mechanics.

The condition $n\lambda^3 = 1$, or

$$n\left(\frac{2\pi\hbar^2}{mkT}\right)^{3/2} = 1 \tag{8.5}$$

defines a line in the $T$–$n$ plane that serves as a rough division between the classical and quantum regimes, as indicated in Fig. 8.1.

The degeneracy temperature $T_0$ is given by

$$kT_0 = \left(\frac{2\pi\hbar^2}{m}\right)n^{2/3}. \tag{8.6}$$

Its value varies over a wide range for different physical systems, as indicated in Table 8.1. For example, at room temperature, a gas at STP can be described classically, whereas electrons in a metal are in the extreme quantum region. Liquid helium has a degeneracy temperature in between. At 2.17 K it makes a transition to a quantum phase that exhibits superfluidity.

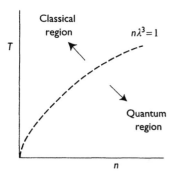

Figure 8.1 Classical and quantum regimes in the temperature–density plane. The rough dividing line is $n\lambda^3 = 1$, where $\lambda$ is the thermal wavelength.

Table 8.1 Quantum degeneracy temperatures

| System | Density (cm³) | $T_0$ (K) |
|---|---|---|
| $H_2$ gas | $2 \times 10^{19}$ | $5 \times 10^{-2}$ |
| Liquid $^4$He | $2 \times 10^{22}$ | 2 |
| Electrons in metal | $10^{22}$ | $10^4$ |

## 8.2 Identical particles

In quantum mechanics atoms are identical, in the sense that the Hamiltonian is invariant under a permutation of their coordinates. This property has no analog in classical mechanics, but it has far-reaching practical consequences.

To illustrate the concept, let us consider two particles with respective coordinates $\mathbf{r}_1, \mathbf{r}_2$. Interchange of the coordinates can be represented by a permutation operation $P$ on the wave function:

$$P\Psi(\mathbf{r}_1, \mathbf{r}_2) = \Psi(\mathbf{r}_2, \mathbf{r}_1). \tag{8.7}$$

Clearly $P^2 = 1$. The Hamiltonian is invariant under the permutation, namely, $PHP^{-1} = H$, This means that the operators $P$ and $H$ commute:

$$[P, H] = 0 \tag{8.8}$$

and we can simultaneously diagonalize them. If $\Psi$ is an eigenfunction of $H$ with energy eigenvalue $E$, then so is $P\Psi$, with the same energy:

$$
\begin{aligned}
H\Psi &= E\Psi, \\
PH\Psi &= EP\Psi, \\
(PHP^{-1})(P\Psi) &= E(P\Psi), \\
H(P\Psi) &= E(P\Psi).
\end{aligned}
\tag{8.9}
$$

If the energy is not degenerate, then $P\Psi$ and $\Psi$ must describe the same state, and $P\Psi$ can differ from $\Psi$ at most by a normalization factor. Since $P^2 = 1$, we can choose that factor to be $\pm 1$. Thus

$$\Psi(\mathbf{r}_1, \mathbf{r}_2) = \pm\Psi(\mathbf{r}_2, \mathbf{r}_1). \tag{8.10}$$

That is, the wave function must be either symmetric or antisymmetric under the interchange of particle coordinates. Particles with the symmetric property are said to obey *Bose statistics*, and are called *bosons*, while those with the antisymmetric property are said to obey *Fermi statistics*, and are called *fermions*. The wave function gives the probability amplitude of finding one particle at $\mathbf{r}_1$ and one particle at $\mathbf{r}_2$, but it cannot tell us which one.

The quantum-mechanical concept of identical particles affects the way we count states. Consider two identical particles localized at separated points $A$ and $B$. Let $f(x)$ and $g(x)$ denote the one-particle wave function localized about $A$ and $B$, respectively, with no overlap between the wave functions, as illustrated in Fig. 8.2. If the particles were distinguishable, we would have two possible states $f(x_1)g(x_2)$ and $f(x_2)g(x_1)$, corresponding to particle 1 at $A$ and particle 2 at $B$, and vice versa.

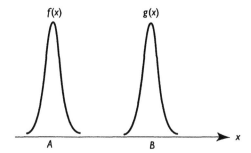

$f(x)$  $g(x)$

$A$  $B$  $x$

*Figure 8.2* We can place two identical particles at the sites $A$ and $B$, but it is impossible to specify which one is at $A$ and which at $B$.

For identical particles, however, there is only one state, with wave function

$$\Psi(x_1, x_2) = f(x_1)g(x_2) \pm f(x_2)g(x_1), \qquad (8.11)$$

where the $+$ sign corresponds to Bose statistics, and the $-$ sign to Fermi statistics. Since indistinguishability affects the counting of states, it has a direct effect on the entropy of the system.

## 8.3 Occupation numbers

For $N$ identical particles, the wave function $\Psi(\mathbf{r}_1, \ldots, \mathbf{r}_N)$ must be an even (odd) function under the interchange of any pair of coordinates for bosons (fermions). We can construct a complete set of $N$-body wave functions by first considering a complete orthonormal set of single-particle wave functions $u_\alpha(\mathbf{r})$, satisfying

$$\int d^3 r\, u_\alpha^*(\mathbf{r}) u_\beta(\mathbf{r}) = \delta_{\alpha\beta}, \qquad (8.12)$$

where $\alpha$ is a single-particle quantum number. We then form a product of $N$ of these wave functions, and symmetrize or antisymmetrize the product with respect to the coordinates $\{\mathbf{r}_1, \ldots, \mathbf{r}_N\}$. Equivalently, we can symmetrize (or antisymmetrize) with respect to the set of single-particle labels $\{\alpha_1, \ldots, \alpha_N\}$. The $N$-body wave function so obtained has the form

$$\Psi(\mathbf{r}_1, \ldots, \mathbf{r}_N) = \frac{1}{\sqrt{N!}} \sum_P \delta_P P \left[ u_{\alpha_1}(\mathbf{r}_1) \cdots u_{\alpha_N}(\mathbf{r}_N) \right], \qquad (8.13)$$

where $P$ is a permutation of the set $\{\alpha_1, \ldots, \alpha_N\}$, and the sum extends over all $N!$ permutations. The signature factor $\delta_P$ is unity for bosons, while for fermions it is given by

$$\delta_P = \begin{cases} 1 & \text{if } P \text{ is an even permutation} \\ -1 & \text{if } P \text{ is an odd permutation} \end{cases} \quad \text{(for fermions)} \tag{8.14}$$

In the fermion case, the wave function is a determinant, called the *Slater determinant*:

$$\Psi(\mathbf{r}_1, \ldots, \mathbf{r}_N) = \frac{1}{\sqrt{N!}} \begin{vmatrix} u_1(\mathbf{r}_1) & u_1(\mathbf{r}_2) & \cdots & u_1(\mathbf{r}_N) \\ u_2(\mathbf{r}_1) & u_2(\mathbf{r}_2) & \cdots & u_2(\mathbf{r}_N) \\ \vdots & \vdots & & \vdots \\ u_N(\mathbf{r}_1) & u_N(\mathbf{r}_2) & \cdots & u_N(\mathbf{r}_N) \end{vmatrix} \quad \text{(for fermions)} \tag{8.15}$$

We see immediately that $\Psi$ vanishes whenever two single-particle wave functions are the same, e.g. if $u_1(\mathbf{r}) \equiv u_2(\mathbf{r})$. This is the statement of the *Pauli exclusion principle*.

Because of indistinguishability, the $N$-body wave function is labeled by the set $\{\alpha_1, \cdots, \alpha_N\}$, in which the ordering of the set is irrelevant. To make this explicit, we specify the number of particles $n_\alpha$ in the state $\alpha$. The number $n_\alpha$ is called the *occupation number* of the single-particle state $\alpha$, with the allowed values

$$n_\alpha = \begin{cases} 0, 1, 2, \cdots, \infty & \text{for bosons,} \\ 0, 1 & \text{for fermions.} \end{cases} \tag{8.16}$$

For an $N$-particle system, they satisfy the condition

$$\sum_\alpha n_\alpha = N. \tag{8.17}$$

A sum over states is a sum over all possible sets $\{n_\alpha\}$ satisfying (8.17).

For free particles, it is convenient to choose the single-particle functions to be plane waves. The label $\alpha$ corresponds to the wave vector $\mathbf{k}$:

$$u_\mathbf{k}(\mathbf{r}) = \frac{1}{\sqrt{V}} e^{i\mathbf{k}\cdot\mathbf{r}}. \tag{8.18}$$

In the thermodynamic limit, we can replace the sum over plane-wave states by an integral:

$$\sum_k \rightarrow \frac{V}{(2\pi)^3} \int d^3k = \int \frac{d^3r\, d^3p}{h^3}. \tag{8.19}$$

This has the form of a classical phase-space integral, with unit specified to be $h^3$, the minimum cell size according to the uncertainty principle. In this limit (8.17) for a uniform gas can be written as

$$\int \frac{d^3p}{h^3} n_{\mathbf{p}} = n, \tag{8.20}$$

where $n_{\mathbf{p}}$ is the number of particles with momentum $\mathbf{p}$, and $n = N/V$ is the density. In the classical limit, $n_{\mathbf{p}}$ is related to the momentum distribution function $f(\mathbf{p})$ defined in (5.17) through

$$n_{\mathbf{p}} = h^3 f(\mathbf{p}). \tag{8.21}$$

## 8.4  Spin

For particles with spin, the single-particle states are labeled by wave vector $\mathbf{k}$ and spin state $s$:

$$\alpha = \{\mathbf{k}, s\}. \tag{8.22}$$

The single-particle wave function is given by

$$u_{\mathbf{k}s}(\mathbf{r}) = \frac{1}{\sqrt{V}} e^{i\mathbf{k}\cdot\mathbf{r}} \chi_s(\sigma), \tag{8.23}$$

where $\sigma$ is the spin coordinate. For a spin $\frac{1}{2}$ particle such as the electron, both $s$ and $\sigma$ are two-valued, and

$$\chi_s(\sigma) = \delta_{s\sigma}. \tag{8.24}$$

Sometimes it is convenient to display the spin components in matrix form:

$$u_{\mathbf{k}}(\mathbf{r}) = \begin{pmatrix} u_{\mathbf{k},+1}(\mathbf{r}) \\ u_{\mathbf{k},-1}(\mathbf{r}) \end{pmatrix}. \tag{8.25}$$

In most of our applications, we will not be concerned with the wave functions. What is important is that we include all quantum numbers in the sum over states. For particles of spin $S$, we have

$$N = \sum_{\alpha} n_{\alpha} = \sum_{s=1}^{2S+1} \sum_{\mathbf{k}} n_{\mathbf{k}s}. \tag{8.26}$$

For a free gas in equilibrium, $n_{\mathbf{k}s}$ is independent of the spin state, and the spin sum simply gives a factor $2S + 1$. Thus, the particles with different spin orientations form independent gases, each containing $N/(2S+1)$ particles. This is why we can speak of spinless fermions, from a mathematical point of view.

## 8.5  Microcanonical ensemble

The microcanonical ensemble is a collection of states with equal weight. For non-interacting particles, states specified by the occupations numbers $\{n_\alpha\}$ are equally probable, as long as they satisfy the constraints

$$\sum_\alpha n_\alpha = N,$$

$$\sum_\alpha \epsilon_\alpha n_\alpha = E, \tag{8.27}$$

where $\epsilon_\alpha$ is the energy of the single-particle state labeled by $\alpha$. In the thermodynamic limit, the spectrum of states form a continuum. As an aid to the counting of states, we divide the spectrum into discrete cells numbered $i = 1, 2, 3, \ldots$, with $g_i$ states in cell $i$, as illustrated schematically in Fig. 8.3. Each cell must contain a large number of states, but the cell should be small enough that the energies of the states are unresolved on a macroscopic scale. We denote the cell energy by $\epsilon_i$, and refer to $g_i$ as the cell degeneracy. The cell occupation numbers $n_i$ satisfy the constraints

$$\sum_i n_i = N,$$

$$\sum_i \epsilon_i n_i = E. \tag{8.28}$$

It should be noted that the division into cells is an imprecise coarse-graining process. The cell degeneracy $g_i$ is an interim number that should disappear from the final expressions of physical quantities.

Because of coarse graining, the cell occupation $\{n_i\}$ corresponds to more than one quantum-mechanical state. The set $\{\bar{n}_i\}$ corresponding to the maximum number of quantum states is called the most probable set, and we assume that it describes thermal equilibrium. The development here bears a formal resemblance to the classical ensemble discussed in Section 5.6, but the fact that we have identical particles changes the rules for counting states; we have to use "quantum statistics".

To find the number of ways in which $\{n_i\}$ can be realized, we assign particles into single-particle quantum states, without paying attention to which particle goes into which cell, since the exchange of particles in different cells does not produce

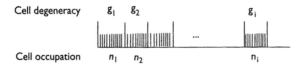

*Figure 8.3* Grouping the continuum of states into cells, in order to facilitate counting.

a new state. Thus, we can independently calculate the occupation of each cell, and multiply the results. The number of ways to obtain $\{n_i\}$ is therefore

$$\Omega\{n_i\} = \prod_j w_j(n_j), \qquad (8.29)$$

where $w_j(n_j)$ is the number of ways to put $n_j$ particles into the $j$th cell, which contains $g_j$ single-particle states.

The formula (8.29) is very different from the corresponding classical formula (5.34). The classical way of counting in effect accepts all wave functions regardless of symmetry properties under the interchange of coordinates. The set of acceptable wave functions is far greater than the union of the two quantum cases. The classical way of counting is sometimes called "Boltzmann statistics", but it actually corresponds to no statistics.

## 8.6 Fermi statistics

For fermions, each single-particle state can accommodate at most one particle, and thus each state is either occupied or empty. To place $n_j$ fermions into $g_j$ states, we pick the $n_j$ occupied states out of a total of $g_j$ states. The number of ways to achieve this is given by the binomial coefficient

$$w_j(n_j) = \frac{g_j!}{n_j!(g_j - n_j)!}. \qquad (8.30)$$

Therefore

$$\Omega\{n_i\} = \prod_j \frac{g_j!}{n_j!(g_j - n_j)!}. \qquad (8.31)$$

Assuming that $n_j$ and $g_j$ are large numbers, we use the Stirling approximation to obtain

$$\ln \Omega\{n_i\} = \sum_j \left[ \ln g_j! - \ln n_j! - \ln(g_j - n_j)! \right]$$

$$\approx \sum_j \left[ g_j \ln g_j - n_j \ln n_j - (g_j - n_j) \ln(g_j - n_j) \right]. \qquad (8.32)$$

The most probable distribution $\{\bar{n}_i\}$ is obtained by maximizing this, subject to constraints. Using Lagrange multipliers, we require

$$\delta \left[ \ln \Omega\{n_i\} + \alpha \sum_j n_j - \beta \sum_j \epsilon_j n_j \right] = 0,$$

under independent variations of each $n_j$. This leads to

$$\sum_j \delta n_j \left[ -\ln n_j + \ln(g_j - n_j) + \alpha - \beta \epsilon_j \right] = 0. \tag{8.33}$$

Since the $\delta n_j$ are independent and arbitrary, we must have

$$-\ln n_j + \ln(g_j - n_j) + \alpha - \beta \epsilon_j = 0, \tag{8.34}$$

which gives

$$\bar{n}_j = \frac{g_j}{e^{-\alpha + \beta \epsilon_j} + 1}. \tag{8.35}$$

This is the Fermi distribution.

## 8.7  Bose statistics

The $j$th cell is subdivided into $g_j$ compartments by means of $g_j - 1$ partitions, and each compartment can hold any number of bosons. We can vary the particle numbers in the compartments by repositioning the partitions. The number of ways to populate the cell with $n_j$ bosons is to throw the $n_j$ particles into the cell, in any manner whatsoever, and then count the number of distinct configurations obtainable, when we permute the $n_j$ particles together with the $g_j - 1$ partitions:

$$w_j(n_j) = \frac{(n_j + g_j - 1)!}{n_j!(g_j - 1)!}. \tag{8.36}$$

This gives

$$\Omega\{n_i\} = \prod_j \frac{(n_j + g_j - 1)!}{n_j!(g_j - 1)!},$$

$$\ln \Omega\{n_i\} \approx \sum_j \left[ (n_j + g_j) \ln(n_j + g_j) - n_j \ln n_j - g_j \ln g_j \right], \tag{8.37}$$

where we have put $g_j - 1 \approx g_j$. The most probable distribution is obtained through the condition

$$\sum_j \delta n_j \left[ \ln(n_j + g_j - 1) - \ln n_j + \alpha - \beta \epsilon_j \right] = 0, \tag{8.38}$$

with the result

$$\bar{n}_j = \frac{g_j}{e^{-\alpha + \beta \epsilon_j} - 1}. \tag{8.39}$$

This is the Bose distribution.

## 8.8  Determining the parameters

The Fermi and Bose cases can be discussed together, by writing

$$n_j = \frac{g_j}{e^{-\alpha+\beta\epsilon_j} \pm 1} \qquad +: \text{Fermi}, \quad -: \text{Bose}, \tag{8.40}$$

where we have omitted the bar over $n_j$ for simplicity. The total number of particles is given by

$$N = \sum_j \frac{g_j}{e^{-\alpha+\beta\epsilon_j} \pm 1}. \tag{8.41}$$

We may replace the sum over cells $j$ by a sum over states $\mathbf{k}$, through the replacement

$$\sum_j g_j \rightarrow \sum_{\mathbf{k}} 1. \tag{8.42}$$

Thus

$$N = \sum_{\mathbf{k}} \frac{1}{e^{-\alpha+\beta\epsilon_k} \pm 1}, \tag{8.43}$$

where $\epsilon_k = \hbar^2 k^2/2m$, with $k = |\mathbf{k}|$. The cell degeneracy, which was a mathematical device, has now disappeared from the formula.

In the thermodynamic limit, we have

$$n = \int \frac{d^3k}{(2\pi)^3} \frac{1}{e^{-\alpha+\beta\epsilon_k} \pm 1} = \frac{1}{2\pi^2} \int_0^\infty dk \, \frac{k^2}{e^{-\alpha+\beta\epsilon_k} \pm 1}, \tag{8.44}$$

where $n$ is the density. Let us make the substitution $\beta\epsilon_k = x^2$, or $k = x\sqrt{2m/\beta\hbar^2}$. Then

$$n = \frac{1}{2\pi^2} \left(\frac{2m}{\beta\hbar^2}\right)^{3/2} \int_0^\infty dx \, \frac{x^2}{e^{-\alpha+x^2} \pm 1}. \tag{8.45}$$

The density should remain finite in the limit $\beta \rightarrow 0$. This requires that the integral vanish, which in turn requires

$$e^{-\alpha} \xrightarrow[\beta\to 0]{} \infty. \tag{8.46}$$

Thus the term $\pm 1$ can be neglected in this limit, and we have

$$n_{\mathbf{k}} \xrightarrow[\beta\to 0]{} e^{\alpha-\beta\epsilon_k}, \tag{8.47}$$

which corresponds to the Maxwell–Boltzmann distribution. Therefore, we define the temperature $T$ through

$$\beta = \frac{1}{kT}. \tag{8.48}$$

Both the Bose and Fermi distributions approach the classical distribution in the high-temperature limit, because at high temperatures the particles tend to populate the excited states sparsely, and when there are few particles in a state, it matters little whether they are bosons or fermions.

The Lagrange multiplier $\alpha$ is determined by the condition (8.45), which can be rewritten as

$$n\lambda^3 = \frac{4}{\sqrt{\pi}} \int_0^\infty dx \, \frac{x^2}{z^{-1}e^{x^2} \pm 1}, \tag{8.49}$$

where $z = e^\alpha$ is called the *fugacity*. The *chemical potential* $\mu$ is defined through

$$z = e^{\beta\mu}. \tag{8.50}$$

The occupation number for state $\mathbf{k}$ can now be written in the form

$$n_\mathbf{k} = \frac{1}{z^{-1}e^{\beta\epsilon_k} \pm 1}. \tag{8.51}$$

This is the most probable occupation number, which is in principle different from the ensemble average. As in the classical case, the two coincide in the thermo-dynamic limit, but we postpone the proof to Chapter 13, where formal tools will be introduced.

## 8.9   Pressure

The pressure of the quantum ideal gas can be calculated using (6.11), which is based on the particle flux impinging on a wall arising from a momentum distribution $f(\mathbf{p})$:

$$P = \int_{v_x>0} d^3p \, (2mv_x)v_x f(\mathbf{p}) = m \int d^3p \, v_x^2 f(\mathbf{p}). \tag{8.52}$$

Using $f(\mathbf{p}) = h^{-3}n_\mathbf{p}$, and the fact that $v_x^2$ can be replaced by $\frac{1}{3}v^2$, we can rewrite

$$P = \frac{2}{3} \int \frac{d^3k}{(2\pi)^3} \frac{\epsilon_k}{z^{-1}e^{\beta\epsilon_k} \pm 1}. \tag{8.53}$$

Changing the variable of integration such that $\beta\epsilon_k = x^2$, i.e.

$$k = \sqrt{\frac{2mkT}{\hbar^2}} x = \frac{\sqrt{4\pi}x}{\lambda}, \tag{8.54}$$

where $\lambda$ is the thermal wavelength, we obtain

$$\frac{P}{kT} = \frac{1}{\lambda^3} \frac{8}{3\sqrt{\pi}} \int_0^\infty dx \, \frac{x^4}{z^{-1}e^{x^2} \pm 1}. \tag{8.55}$$

The internal energy is given by

$$U = \sum_{\mathbf{k}} \frac{\epsilon_k}{z^{-1}e^{\beta\epsilon_k} \pm 1} = V \int \frac{d^3k}{(2\pi)^3} \frac{\epsilon_k}{z^{-1}e^{\beta\epsilon_k} \pm 1}. \tag{8.56}$$

Comparison with (8.53) yields the relation

$$PV = \tfrac{2}{3}U. \tag{8.57}$$

This holds for the ideal Fermi and Bose gas, as well as the classical ideal gas. It depends only on the fact that the particles move in three dimensions with energy–momentum relation $\epsilon \propto (p_x^2 + p_y^2 + p_z^2)$.

## 8.10 Entropy

The entropy can be obtained from $S = k \ln \Omega$. We use (8.32) for the Fermi case, and (8.37) for the Bose case, to obtain

$$\frac{S}{k} = \sum_{\mathbf{k}} \left[ \frac{1}{\xi \pm 1} \ln(\xi \pm 1) \pm \frac{\xi}{\xi \pm 1} \ln \frac{\xi \pm 1}{\xi} \right], \tag{8.58}$$

where $\xi \equiv z^{-1}e^{-\beta\epsilon}$. Using the definition (8.56) for $U$, and (8.17) for $N$, we obtain

$$\frac{S}{k} = -N \ln z + \frac{U}{kT} \pm \sum_{\mathbf{k}} \ln\left(1 \pm ze^{-\beta\epsilon}\right), \tag{8.59}$$

with + for the Fermi case, and − for the Bose case.

In the thermodynamic limit the last term becomes an integral, and we can make a partial integration. We show this explicitly for the Fermi case:

$$\frac{V}{2\pi^2} \int_0^\infty dk\, k^2 \ln\left(1 + ze^{-\beta\epsilon}\right)$$

$$= \frac{V}{2\pi^2} \left[ \frac{k^3}{3} \ln\left(1 + ze^{-\beta\epsilon}\right) \Big|_0^\infty - \int_0^\infty dk\, \frac{k^3}{3} \frac{\partial}{\partial k} \ln\left(1 + ze^{-\beta\epsilon}\right) \right]$$

$$= \frac{V}{6\pi^2} \frac{\beta\hbar^2}{m} \int_0^\infty dk\, \frac{k^4}{z^{-1}e^{\beta\epsilon} + 1} = \frac{2}{3} \frac{U}{kT} = \frac{PV}{kT}. \tag{8.60}$$

The same final result holds for the Bose case. Therefore, we have the relation

$$S = \frac{1}{T}(U + PV - N\mu), \tag{8.61}$$

where $\mu$ is the chemical potential defined in (8.50). This verifies the thermodynamic relation $T^{-1} = \partial S / \partial U$.

## 8.11   Free energy

We can identify the free energy as

$$A = N\mu - PV, \tag{8.62}$$

for this obeys the Maxwell relations

$$\left(\frac{\partial A}{\partial V}\right)_{V,N} = -P,$$

$$\left(\frac{\partial A}{\partial V}\right)_{V,T} = -\mu. \tag{8.63}$$

Thus (8.61) reduces to the thermodynamic relation

$$A = U - TS. \tag{8.64}$$

From (8.59) we obtain explicitly

$$A = NkT \ln z \mp kT \sum_{\mathbf{k}} \ln \left(1 \pm z e^{-\beta\epsilon}\right), \tag{8.65}$$

where $+$ holds for the Fermi case and $-$ for the Bose case.

## 8.12   Equation of state

The equation of state is no longer the simple relation $P = nkT$. It can be presented in parametric form by combining (8.55) and (8.49):

$$\frac{\lambda^3 P}{kT} = \frac{8}{3\sqrt{\pi}} \int_0^\infty dx \, \frac{x^4}{z^{-1}e^{x^2} \pm 1},$$

$$n\lambda^3 = \frac{4}{\sqrt{\pi}} \int_0^\infty dx \, \frac{x^2}{z^{-1}e^{x^2} \pm 1}. \tag{8.66}$$

The integrals on the right-hand sides can be expanded in powers of $z$:

$$\frac{8}{3\sqrt{\pi}} \int_0^\infty dx \, \frac{x^4}{z^{-1}e^{x^2} \pm 1} = z \mp \frac{z^2}{2^{5/2}} + \frac{z^3}{3^{5/2}} \mp \cdots,$$

$$\frac{4}{\sqrt{\pi}} \int_0^\infty dx \frac{x^2}{z^{-1}e^{x^2} \pm 1} = z \mp \frac{z^2}{2^{3/2}} + \frac{z^3}{3^{3/2}} \mp \cdots \tag{8.67}$$

It is convenient to introduce a class of Fermi functions $f_k(z)$ and Bose functions $g_k(z)$ :

$$f_k(z) \equiv \sum_{\ell=1}^{\infty} (-1)^{\ell+1} \frac{z^\ell}{\ell^k},$$

$$g_k(z) \equiv \sum_{\ell=1}^{\infty} \frac{z^\ell}{\ell^k}. \tag{8.68}$$

They obey the recursion relations

$$z\frac{d}{dz}f_k(z) = -f_{k-1}(z),$$

$$z\frac{d}{dz}g_k(z) = g_{k-1}(z), \tag{8.69}$$

with integral representations

$$\left.\begin{array}{c} f_k(z) \\ g_k(z) \end{array}\right\} = \frac{2^{2k-1}}{\sqrt{\pi}} \frac{\Gamma(k-\frac{1}{2})}{\Gamma(2k-1)} \int_0^\infty dx \frac{x^{2k-1}}{z^{-1}e^{x^2} \pm 1}, \tag{8.70}$$

where $\Gamma(n)$ is the gamma function. We can then write the parametric equations of state for the Fermi case as

$$\frac{\lambda^3 P}{kT} = f_{5/2}(z),$$

$$\lambda^3 n = f_{3/2}(z); \tag{8.71}$$

and the Bose case as

$$\frac{\lambda^3 P}{kT} = g_{5/2}(z),$$

$$\lambda^3 n = g_{3/2}(z). \tag{8.72}$$

The Fermi functions have a singularity at $z = -1$, which lies outside of the physical region. The Bose functions have a singularity at $z = 1$, which leads to Bose–Einstein condensation, as we shall discuss in a later chapter.

## 8.13  Classical limit

In the high-temperature limit $\lambda \to 0$ and hence $z \to 0$, both quantum gases approach the classical behavior $P = nkT$. As the temperature decreases, their behaviors begin to diverge from each other, and when we approach the quantum region $n\lambda^3 \approx 1$ they become dramatically different. We examine the classical neighborhood in more detail.

We want to solve for the fugacity as a function of density and temperature from the equation

$$n\lambda^3 = z \mp \frac{z^2}{2^{3/2}} + \frac{z^3}{3^{3/2}} \mp \cdots \qquad (8.73)$$

In the first approximation, we take

$$z \approx n\lambda^3. \qquad (8.74)$$

The occupation number approaches the Maxwell–Boltzmann distribution:

$$n_{\mathbf{p}} \approx h^3 f(\mathbf{p}) \qquad (8.75)$$

and, as noted before, we recover the classical equation of state.

The next approximation can be obtained by iteration. Rewrite (8.73) to second order in the form

$$z = n\lambda^3 \pm \frac{z^2}{2^{3/2}} \qquad (8.76)$$

and substitute the first approximation $z = n\lambda^3$ on the right-hand side. This gives

$$z = n\lambda^3 \left( 1 \pm \frac{n\lambda^3}{2^{3/2}} + \cdots \right),$$

$$\mu = kT \left[ \ln \left( n\lambda^3 \right) \pm \frac{n\lambda^3}{2^{3/2}} + \cdots \right]. \qquad (8.77)$$

The second term is the first quantum correction, which has opposite signs for Fermi and Bose statistics. The equations of state are

$$\frac{\lambda^3 P}{kT} = z \mp \frac{z^2}{2^{5/2}} + \cdots$$

$$= \left( n\lambda^3 \pm \frac{\left( n\lambda^3 \right)^2}{2^{3/2}} \right) \mp \frac{1}{2^{5/2}} \left( n\lambda^3 \pm \frac{\left( n\lambda^3 \right)^2}{2^{3/2}} \right)^2 + \cdots \qquad (8.78)$$

which leads to

$$\frac{P}{nkT} = 1 \pm 2^{-5/2} n\lambda^3 + \cdots \qquad (8.79)$$

where the upper signs are for Fermi statistics and the lower signs for Bose statistics.

Compared to the classical gas of the same density and temperature, the pressure is larger for Fermi statistics, and smaller for Bose statistics. This indicates an effective repulsion between identical fermions, and attraction between identical bosons, even though there is no interparticle potential. These effective interactions arise from the correlations imposed by the symmetry of the quantum-mechanical wave function.

## Problems

**8.1**  The thermal wavelength $\lambda = \sqrt{2\pi\hbar^2/mkT}$ applies only to non-relativistic particles. For an ultra-relativistic particle, or a photon, show that it would be replaced by

$$L = \frac{\hbar c}{kT},$$

where $c$ is the velocity of light. What would be the degeneracy temperature $T_0$?

**8.2**  Consider the temperature–density ($T$–$n$) diagram of a free electron gas. Indicate the regions in which the system should be treated (a) relativistically, and (b) quantum mechanically.

**8.3**  At high temperatures the heat capacity $C_V$ of a non-relativistic monatomic gas approaches $\frac{3}{2}Nk$. Find the first quantum correction for both Fermi and Bose statistics. Express the correction in terms of $T/T_0$, where $T_0$ is the degeneracy temperature.

**8.4**  For a gas at very low density or very high temperature, such that $\lambda^3 n \rightarrow 0$, the occupation number approaches

$$n_\lambda = z e^{-\beta\epsilon_\lambda},$$

independent of statistics. Although this has classical form, the energy spectrum $\epsilon_\lambda$ is still quantum mechanical.

(a)  Show that $z = N/Q$, where

$$Q = \sum_\lambda e^{-\beta\epsilon_\lambda}$$

is called the partition function.

(b)  Show that the internal energy per particle is given by

$$\frac{U}{N} = -\frac{\partial}{\partial\beta}\ln Q$$

(c)  For a polyatomic molecule, the energy has contributions from translational motion, and internal degrees of freedom such as rotations and vibrations: $\epsilon = \epsilon_{\text{trans}} + \epsilon_{\text{rot}} + \epsilon_{\text{vib}}$. Show that the partition function factorizes, and the specific heat capacity decomposes into a sum of terms:

$$Q = Q_{\text{trans}} Q_{\text{rot}} Q_{\text{vib}},$$

$$c_V = c_{\text{trans}} + c_{\text{rot}} + c_{\text{vib}},$$

where the subscripts refer to contributions from translational, rotational, and vibrational modes.

**8.5**   The translational energy is labeled by a wavevector $\mathbf{k}$, with $\epsilon_{\mathbf{k}} = \hbar^2 k^2/2m$. Show that

$$Q_{\text{trans}} = \frac{V}{\lambda^3},$$

$$\frac{c_{\text{trans}}}{k} = \frac{3}{2}.$$

**8.6**   A simple model for the rotational energy is $\epsilon_{\ell m} = \hbar^2 \ell(\ell + 1)/2I$, where $\ell = 0, 1, 2, \ldots; m = -\ell, \ldots, \ell;$ and $I$ is the moment of inertia. Thus

$$Q_{\text{rot}} = \sum_{\ell=0}^{\infty} (2\ell + 1)\, e^{-\beta \hbar^2 \ell(\ell+1)/2I}.$$

(a)   For $kT \ll \hbar^2/2I$ keep only the first two terms in $Q_{\text{rot}}$. Show that

$$\frac{c_{\text{rot}}}{k} \approx 3 \left( \frac{\beta \hbar^2}{I} \right)^2 e^{-\beta \hbar^2/I}.$$

(b)   For $kT \gg \hbar^2/2I$ approximate the sum over $\ell$ by an integral. Show that

$$\frac{c_{\text{rot}}}{k} \approx 1.$$

(c)   Make a qualitative sketch of $U/N$ and $c_r$ as functions of temperature. Does $c_{\text{rot}}$ approach the asymptotic value from above or from below?

**8.7**   The vibrational energy is $\epsilon_n = \hbar\omega \left( n + \frac{1}{2} \right)$, where $n = 0, 1, 2, \ldots$ and $\omega$ is the vibrational frequency.

(a)   Show that

$$\frac{c_{\text{vib}}}{k} = e^{-\beta \hbar\omega} \left( \frac{\beta \hbar\omega}{1 - e^{-\beta \hbar\omega}} \right)^2.$$

Make a qualitative sketch of $c_{\text{vib}}$ as a function of temperature.

(b)   Find the mean value $\langle n + \frac{1}{2} \rangle$ and mean-square fluctuation $\langle (n + \frac{1}{2})^2 \rangle - \langle n + \frac{1}{2} \rangle^2$.

**8.8**   On the basis of the specific heats calculated above, we can now understand why equipartition works only among degrees of freedom excited, as illustrated in Fig. 6.2. Reproduce that figure qualitatively, and relate the threshold temperatures to the parameters in the energy spectrum.

**8.9**   Consider anharmonic vibrations with energy $\epsilon_n = \hbar\omega[(n + \frac{1}{2}) + b(n + \frac{1}{2})^2]$, where $b \ll 1$. Find the vibrational specific heat to first order in $b$. Use the results of Problem 8.7(b).

# Chapter 9

# The Fermi gas

## 9.1  Fermi energy

In the low-temperature limit $T \to 0$, the Fermi distribution has the behavior

$$n_{\mathbf{k}} = \frac{1}{e^{\beta(\epsilon_k - \mu)} + 1} \xrightarrow[T \to 0]{} \begin{cases} 0 & \text{if } \epsilon_k > \mu, \\ 1 & \text{if } \epsilon_k < \mu. \end{cases} \tag{9.1}$$

Using the step function

$$\theta(x) = \begin{cases} 1 & \text{if } x > 0, \\ 0 & \text{if } x < 0, \end{cases} \tag{9.2}$$

we can write

$$n_{\mathbf{k}} \xrightarrow[T \to 0]{} \theta(\mu - \epsilon_k). \tag{9.3}$$

This means that all states with energy below the *Fermi energy*

$$\epsilon_{\mathrm{F}} = \mu(n, 0) \tag{9.4}$$

are occupied, and all those above are empty. In momentum space the occupied states lie within the Fermi sphere of radius $p_F$, as illustrated in Fig. 9.1. Such a condition is called "quantum degeneracy".

We can determine the Fermi energy through the condition

$$N = \sum_{\substack{\text{states with} \\ \epsilon < \epsilon_{\mathrm{F}}}} 1. \tag{9.5}$$

For free fermions of spin $S$ the condition gives

$$N = \frac{V(2S + 1)}{(2\pi)^3} \frac{4\pi}{3} k_{\mathrm{F}}^3, \tag{9.6}$$

where $k_{\mathrm{F}} = p_{\mathrm{F}}/\hbar$ is the Fermi wave number. Thus

$$n = \frac{(2S + 1)k_{\mathrm{F}}^3}{6\pi^2}, \tag{9.7}$$

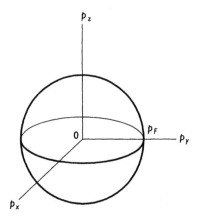

Figure 9.1  At absolute zero, all states within the Fermi sphere in momentum are occupied, and all others are empty. The radius of the sphere is the Fermi momentum $p_F = \hbar k_F$, where $k_F$ is the Fermi wave vector. The Fermi energy is $\epsilon_F = p_F^2/2m$.

and

$$
k_F = \left( \frac{6\pi^2 n}{2S+1} \right)^{1/3},
$$

$$
\epsilon_F = \frac{\hbar^2}{2m} \left( \frac{6\pi^2 n}{2S+1} \right)^{2/3}.
$$

(9.8)

## 9.2   Ground state

At absolute zero, the system is in its quantum-mechanical ground state. The internal energy is given by

$$
U_0 = \sum_{|k|<k_F} \epsilon_k = \frac{V(2S+1)}{(2\pi)^3} \int_0^{k_F} dk(4\pi k^2) \frac{\hbar^2 k^2}{2m}
$$

$$
= \frac{V}{(2\pi)^3} \left( \frac{\hbar^2}{2m} \right) \frac{4\pi}{5} k_F^5.
$$

(9.9)

On the other hand,

$$
N = \frac{V(2S+1)}{(2\pi)^3} \frac{4\pi}{3} k_F^3.
$$

(9.10)

Thus, the internal energy per particle at absolute zero is

$$
\frac{U_0}{N} = \frac{3}{5} \frac{\hbar^2 k_F^2}{2m} = \frac{3}{5}\epsilon_F.
$$

(9.11)

This relation is independent of $S$.

We have shown in Section 8.9 that $PV = \frac{2}{3}U$ at any temperature. Thus, at absolute zero

$$P_0 = \frac{2}{5}n\epsilon_F. \tag{9.12}$$

This zero-point pressure arises from the fact that there must be moving particles at absolute zero, since the zero-momentum state can hold only one particle of given spin state. Assuming a metal to be a Fermi gas of electrons contained in a box, with a density $n \approx 10^{22}\,\mathrm{cm}^{-3}$, we find a zero-point pressure

$$P_0 \approx 10^{10}\,\mathrm{erg\,cm}^{-3} \approx 10^4\,\mathrm{atm}. \tag{9.13}$$

## 9.3  Fermi temperature

At a finite temperature, the occupation number $n(\epsilon)$ as a function of the single-particle energy $\epsilon$ has the qualitative shape shown in the left panel in Fig. 9.2. In the right panel, we show the momentum space, where particles within a layer beneath the Fermi surface are excited to a layer above, leaving "holes" beneath the Fermi surface.

We define the Fermi temperature $T_F$ by

$$\epsilon_F = kT_F. \tag{9.14}$$

For low temperatures $T \ll T_F$, the distribution deviates from that at $T = 0$ mainly in the neighborhood of $\epsilon = \epsilon_F$, in a layer of the thickness of order $kT$. That is, particles at energies of order $kT$ below the Fermi energy are excited to energies of order $kT$ above the Fermi energy. For $T \gg T_F$ the distribution approaches the classical Maxwell–Boltzmann distribution.

The Fermi energy for electrons in a metal is typically $\epsilon_F \approx 2\,\mathrm{eV}$, which corresponds to $T_F \approx 2 \times 10^4\,\mathrm{K}$. Therefore, at room temperatures electrons are frozen

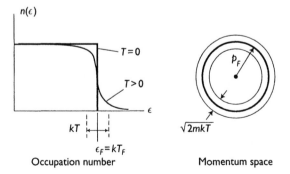

Figure 9.2 Left panel: Occupation number of a Fermi gas near absolute zero. Right panel: A layer beneath the Fermi surface is excited to above the Fermi surface, leaving holes behind.

below the Fermi level, except for a fraction $T/T_F \approx 0.015$. Since the average excitation energy per particle is $kT$, the internal energy is of order $(T/T_F)\,NkT$, and the specific heat capacity tends to zero:

$$\frac{C}{Nk} \sim \frac{T}{T_F}. \tag{9.15}$$

This is why the electronic contributions to the specific heat of a metal can be neglected at room temperature.

## 9.4  Low-temperature properties

The large $z$ limit corresponds to $n\lambda^3 \gg 1$. At a fixed density, this corresponds to $T \ll T_F$, the degenerate limit. The condition for $z$ is given in Section 8.12:

$$n\lambda^3 = f_{3/2}(z). \tag{9.16}$$

In the low-temperature region $z$ is large, and we cannot use the power series for $f_{3/2}$. The relevant asymptotic formula is derived in the Appendix:

$$f_{3/2}(z) \approx \frac{4}{3\sqrt{\pi}}\left[(\ln z)^{3/2} + \frac{\pi^2}{8}\frac{1}{\sqrt{\ln z}} + \cdots\right]. \tag{9.17}$$

Keeping only the first term above, we obtain from (9.16)

$$\ln z \approx \left(\frac{3\sqrt{\pi}}{4}n\lambda^3\right)^{2/3} = \frac{T_F}{T}. \tag{9.18}$$

This gives the correct limiting value for the chemical potential $\mu = \epsilon_F$.

To the next order, we use the equation

$$n\lambda^3 \approx \frac{4}{3\sqrt{\pi}}\left[(\ln z)^{3/2} + \frac{\pi^2}{8}\frac{1}{\sqrt{\ln z}}\right] \tag{9.19}$$

and rewrite it in the form

$$(\ln z)^{3/2} \approx \frac{3\sqrt{\pi}}{4}n\lambda^3 - \frac{\pi^2}{8}\frac{1}{\sqrt{\ln z}}. \tag{9.20}$$

Substituting the first approximation $\ln z = T_F/T$ on the right-hand side, we then obtain

$$\ln z \approx \frac{T_F}{T}\left[1 - \frac{\pi^2}{12}\left(\frac{T}{T_F}\right)^2\right],$$

$$\mu \approx \epsilon_F\left[1 - \frac{\pi^2}{12}\left(\frac{T}{T_F}\right)^2\right]. \tag{9.21}$$

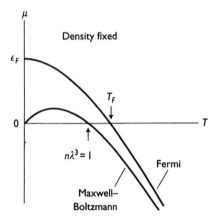

*Figure 9.3* Chemical potential of the ideal Fermi gas. The Maxwell–Boltzmann case is shown for comparison.

Combining this with the high-temperature limit given in Section 8.13, we can sketch the qualitative behavior of the chemical potential as a function of temperature, as shown in Fig. 9.3, where the Maxwell–Boltzmann curve refers to the high-temperature limit $\mu = kT \ln(n\lambda^3)$.

The internal energy is given by

$$U = \sum_{\mathbf{k}} \epsilon_{\mathbf{k}} n_{\mathbf{k}} = \frac{V}{(2\pi)^3} \frac{4\pi\hbar^2}{2m} \int_0^{\infty} dk\, k^4 n_{\mathbf{k}}. \tag{9.22}$$

We make a partial integration:

$$
\begin{aligned}
U &= -\frac{V}{(2\pi)^3} \frac{4\pi\hbar^2}{2m} \int_0^{\infty} dk \frac{k^5}{5} \frac{\partial n_{\mathbf{k}}}{\partial k} \\
&= \frac{V}{(2\pi)^3} \frac{4\pi\hbar^2 \beta}{2m} \int_0^{\infty} dk \frac{k^5}{5} \frac{z^{-1}e^{\beta\epsilon}}{\left[z^{-1}e^{\beta\epsilon} + 1\right]^2} \frac{\partial\epsilon}{\partial k} \\
&= \frac{V\beta\hbar^4}{20\pi^2 m^2} \int_0^{\infty} dk \frac{k^6 e^{\beta(\epsilon-\mu)}}{\left[e^{\beta(\epsilon-\mu)} + 1\right]^2},
\end{aligned}
\tag{9.23}
$$

where $\epsilon = \hbar^2 k^2/2m$. At low temperatures the integrand is peaked about $k = k_{\mathrm{F}}$. We evaluate it by expanding $k^6$ about that point, and use the expansion for $\mu$ obtained earlier. The result is

$$U = \frac{3}{5} N\epsilon_{\mathrm{F}} \left[1 + \frac{5\pi^2}{12} \left(\frac{T}{T_{\mathrm{F}}}\right)^2 + \cdots\right], \tag{9.24}$$

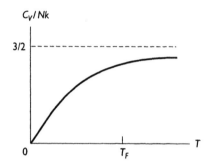

Figure 9.4 Specific heat of ideal Fermi gas.

from which we obtain the equation of state

$$P = \frac{2}{3}\frac{U}{V} = \frac{2}{5}n\epsilon_F\left[1 + \frac{5\pi^2}{12}\left(\frac{T}{T_F}\right)^2 + \cdots\right].$$   (9.25)

The heat capacity is given by

$$\frac{C_V}{Nk} = \frac{\pi^2}{2}\frac{T}{T_F} + \cdots$$   (9.26)

A qualitative plot of the specific heat is shown in Fig. 9.4.

## 9.5   Particles and holes

The absence of a fermion of energy $\epsilon$, momentum $\mathbf{p}$, charge $e$, corresponds to the presence of a hole with

$$\text{Energy} = -\epsilon,$$

$$\text{Momentum} = -\mathbf{p},$$

$$\text{Charge} = -e.$$   (9.27)

Since the number of fermions in a quantum state is 0 or 1, the number of holes is 1 minus the number of fermions. It follows that, in a Fermi gas in thermal equilibrium, the average occupation number for particles and holes are given by

$$\text{Particle:}\quad n_{\mathbf{p}} = \frac{1}{e^{\beta(\epsilon-\mu)} + 1},$$

$$\text{Hole:}\quad 1 - n_{\mathbf{p}} = \frac{1}{e^{\beta(\mu-\epsilon)} + 1},$$   (9.28)

where $\epsilon = \mathbf{p}^2/2m$. As illustrated in Fig. 9.5, the number of excited particles is equal to the number of holes. When an excited fermion falls into an unoccupied

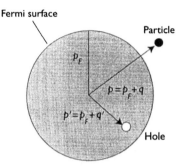

*Figure 9.5* Particle excited to above the Fermi surface, leaving a hole beneath the Fermi surface.

state, it appears that a particles and a hole have annihilated each other. In this sense, a hole is an "antiparticle".

The concept of holes is useful only at low temperatures $T \ll T_F$, when there are relatively few of them below the Fermi surface, with the same number of excited particles above the Fermi surface. Only the holes and excited particles participate in thermal activities, while the rest of the fermions, which constitute the overwhelming majority, lie dormant beneath the Fermi surface. When the temperature is so high that there are a large number of holes, the Fermi surface is "washed out", and the system approaches a Maxwell–Boltzmann distribution of particles.

## 9.6  Electrons in solids

An electron in the ionic lattice of a solid sees a periodic potential, and experiences partial transmission and reflection as it travels through the lattice. The reflected waves can interfere coherently with the original wave to produce complete destructive interference in certain energy ranges. Thus, energy gaps occur in the spectrum of the electron, and the lattice acts like a band-pass filter. A two-band spectrum is illustrated in Fig. 9.6, with a "valence band" with maximum energy $\epsilon_0$, separated by an energy gap $\Delta$ from the "conduction band" above it. Suppose the energy bands are filled with electrons. The two spin states of the electron only supply a factor of 2 in the counting of states. If the density is very low, we have a free Fermi gas with $\epsilon_F \ll \epsilon_0$, and the band structure has little relevance. Interesting phenomena happen when the valence band is almost filled, or a bit overfilled.

Suppose at absolute zero the valence band is completely filled, and the conduction band completely empty. Such a system is called a "natural semiconductor". As the temperature increases, thermal excitation will cause some electrons to be excited across the gap into the conduction band, leaving holes in the valence band.

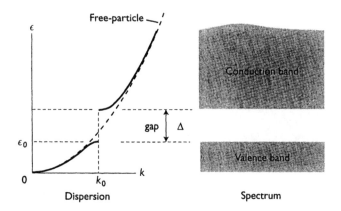

Figure 9.6 Band structure in the energy of an electron in a periodic potential.

The particles and holes behave like free charge carriers, and the system exhibits electrical conductivity. As illustrated in Fig. 9.6, the dispersion curve near the edges of the bands can be locally approximated by parabolas. An excited electron near the bottom of the conduction band has energy

$$\epsilon_c = \epsilon_0 + \Delta + \frac{\mathbf{p}^2}{2m_c}, \tag{9.29}$$

where $\mathbf{p}$ is an effective momentum vector, and $m_c$ is an effective mass. Similarly, an empty state, just below the top of the valence band, has energy

$$\epsilon_v = \epsilon_0 - \frac{\mathbf{p}^2}{2m_v}. \tag{9.30}$$

The density of particles and holes are thus given by

$$n_{\text{part}} = 2 \int \frac{d^3p}{h^3} \frac{1}{e^{\beta(\epsilon_c - \mu)} + 1},$$

$$n_{\text{hole}} = 2 \int \frac{d^3p}{h^3} \frac{1}{e^{\beta(\mu - \epsilon_v)} + 1}. \tag{9.31}$$

Assuming $\beta(\epsilon_c - \mu) \gg 1$, and $\beta(\mu - \epsilon_v) \gg 1$, we have

$$n_{\text{part}} \approx 2 \int \frac{d^3p}{h^3} e^{-\beta(\epsilon_c - \mu)} = 2e^{-\beta(\epsilon_0 + \Delta - \mu)} \left( \frac{m_c kT}{2\pi\hbar^2} \right)^{3/2},$$

$$n_{\text{hole}} \approx 2 \int \frac{d^3p}{h^3} e^{-\beta(\mu - \epsilon_v)} = 2e^{-\beta(\mu - \epsilon_0)} \left( \frac{m_v kT}{2\pi\hbar^2} \right)^{3/2}. \tag{9.32}$$

Overall electrical neutrality requires $n_{\text{part}} = n_{\text{hole}}$, which determines the chemical potential:

$$\mu = \epsilon_0 + \frac{\Delta}{2} + \frac{3}{4}kT \ln \frac{m_c}{m_v}. \tag{9.33}$$

The value at $T = 0$ gives the Fermi energy as $\epsilon_0 + \Delta/2$, which is located in the middle of the energy gap. The approximations made here are valid if $\beta\Delta \gg 1$. The particle and hole densities are now given by

$$n_{\text{part}} = n_{\text{hole}} = 2e^{-\beta\Delta/2} \left( \frac{\sqrt{m_c m_v} kT}{2\pi\hbar^2} \right)^{3/2}. \tag{9.34}$$

For an estimate, use the typical values $\Delta/k = 0.7\,\text{eV}$, and $m_c = m_v = m$, where $m$ is the free electron mass. At room temperature, $T = 300\,\text{K}$, we find $n_{\text{part}} \approx 1.6 \times 10^{13}\,\text{cm}^{-3}$, which is to be compared with a charge carrier density of $10^{20}$ for a metal. The electrical conductivity of a natural semiconductor is therefore negligible.

There are, however, gapless natural semiconductors with $\Delta = 0$, such as $\alpha$-Sn and Hg–Te. In these cases the charge carrier densities become large, and the dependence on temperature goes like $T^{3/2}$.

## 9.7 Semiconductors

The electronic energy spectrum in a real solid rarely looks like that shown in Fig. 9.6, because there always exist impurities in the lattice that can trap electrons into bound states. These impurities act as sources or sinks for electrons, and cases of practical interest are depicted in Fig. 9.7.

In *n-type semiconductors* (*n* for negative), the bound levels lie below the bottom of the conduction band by an energy $\delta \ll \Delta$, and they are filled by electrons at absolute zero. As the temperature increases, these levels donate electrons to the originally empty conduction band. For this reason they are called *donor levels*.

In *p-type semiconductors* (*p* for positive), the bound levels lie above the top of the valence band by $\delta' \ll \Delta$, and are empty at absolute zero. As the temperature increases, they accept electrons excited from the valence band, thereby creating holes. For this reason they are called *acceptor levels*. We neglect excitations between bands.

For the *n*-type semiconductor, let us measure energy with respect to the bottom of the conduction band. Suppose there are $n_D$ bound states per unit volume, of

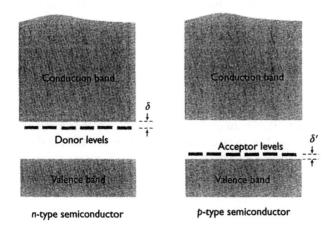

Figure 9.7 Impurities in the lattice of the solid can trap electrons into bound states, with energy levels in the band gap.

energy $-\delta$. The density of electrons in the donor levels at temperature $T$ is

$$n_{\text{donor}} = \frac{2n_D}{e^{-\beta(\delta+\mu)} + 1}.$$
(9.35)

The density of electrons in the conduction band is given by $n_{\text{part}}$ in (9.32), with $\epsilon_0 + \Delta = 0$ in our convention:

$$n_{\text{part}} = \frac{2z}{\lambda^3},$$

$$\lambda = \sqrt{\frac{2\pi\hbar^2}{m_c kT}},$$
(9.36)

where $z = e^{\beta\mu}$ is the fugacity, and $\lambda$ is the thermal wavelength with respect to the effective mass $m_c$. Since electrons in the conduction band were excited from the fully occupied donor levels, we must have $n_{\text{part}} + n_{\text{donor}} = 2n_D$, or

$$\frac{2z}{\lambda^3} + \frac{2n_D}{z^{-1}e^{-\beta\delta} + 1} = 2n_D.$$
(9.37)

To solve for $z$, rewrite this in the form

$$e^{\beta\delta}z^2 + z - n_D\lambda^3 = 0.$$
(9.38)

The solution is

$$z = \tfrac{1}{2}e^{-\beta\delta}\left(\sqrt{4n_D\lambda^3 + 1} - 1\right),$$
(9.39)

which leads to

$$n_{part} = \frac{e^{-\beta\delta}}{\lambda^3} \left( \sqrt{4n_D\lambda^3 + 1} - 1 \right). \tag{9.40}$$

As the temperature rises from absolute zero, the density of particles in the conduction band initially increases exponentially like $T^{3/4}e^{-\delta/kT}$, but flattens to a plateau at $2n_D$, when the donor levels are depleted. The particle density can be adjusted by changing $n_D$. In practice this is done through "doping" – mixing in varying amounts of impurity material.

The p-type semiconductors can be discussed in the same manner, with holes replacing particles:

$$n_{hole} = \frac{e^{-\beta\delta'}}{\lambda^3} \left( \sqrt{4n_A\lambda^3 + 1} - 1 \right), \tag{9.41}$$

where $n_A$ is the number of acceptor states per unit volume and $\lambda = \sqrt{2\pi\hbar^2/m_vkT}$.

## Problems

**9.1** A metal contains a high density of electrons, with interparticle distance of the order of 1 Å. However, the mean free path of electrons at room temperature is very large, of the order of $10^4$ Å. This is because only a small fraction of electrons near the Fermi surface are excited. How does the mean free path depend on the temperature?

**9.2** Model a heavy nucleus of mass number $A$ as a free Fermi gas of an equal number of protons and neutrons, contained in a sphere of radius $R = r_0A^{1/3}$, where $r_0 = 1.4 \times 10^{-13}$ cm. Calculate the Fermi energy and the average energy per nucleon in MeV.

**9.3** Consider a relativistic gas of $N$ particles of spin $\frac{1}{2}$ obeying Fermi statistics, enclosed in volume $V$, at absolute zero. The energy–momentum relation is $E = \sqrt{(pc)^2 + (mc^2)^2}$, where $m$ is the rest mass.
(a) Find the Fermi energy at density $n$.
(b) Define the internal energy $U$ as the average of $E - mc^2$, and the pressure $P$ as the average force per unit area exerted on a perfectly-reflecting wall of the container. Set up expressions for these quantities in the form of integrals, but you need not evaluate them.
(c) Show that $PV = 2U/3$ at low densities, and $PV = U/3$ at high densities. State the criteria for low and high densities.

(d) There may exist a gas of neutrinos (and/or antineutrinos) in the cosmos. (Neutrinos are massless fermions of spin $\frac{1}{2}$.) Calculate the Fermi energy (in eV) of such a gas, assuming a density of one particle per cm$^3$.

**9.4** Consider an ideal gas of non-relativistic atoms at density $n$, with mass $m$, spin $\frac{1}{2}$ and magnetic moment $\mu$. The spin-up and spin-down atoms constitute two independent gases.

(a) In external magnetic field $H$ at absolute zero, the atoms occupy all energy levels below a certain Fermi energy $\epsilon(H)$. Find the number $N_\pm$ of spin-up and spin-down atoms. (*Hint*: The energy of an atom is $(p^2/2m) \mp \mu H$.)
(b) Find the minimum external field that will completely polarize the gas, as a function of the total density $n$.

**9.5** Model a neutron star as an ideal Fermi gas of neutrons at absolute zero, in the gravitational field of a heavy center of mass $M$.

(a) Show that the pressure $P$ of a gas in the gravitational field of a heavy mass $M$ obeys the equation $dP/dr = -\gamma M\rho(r)/r^2$, where $\gamma$ is the gravitational constant, $r$ is the distance from the mass, and $\rho(r)$ is the mass density of the gas.
(b) Show that $P = a\rho^{5/3}$, where $a$ is a constant, and find $\rho$ as a function of distance from the center. (*Hint*: Consider how $P$ and $\rho$ depend on the chemical potential in a Fermi gas.)

**9.6** Consider an ideal Fermi gas of $N$ spinless electrons of mass $m$. In addition to the usual free-particle states, a single electron has $N$ bound states that have the same energy $-\varepsilon$.

(a) Write down expressions for $N_b$ and $N_f$, the average number of bound and free particles in the gas.
(b) Write down the condition determining the fugacity $z = e^{\beta\mu}$.
(c) Find $z$ as a function of temperature and density, assuming that $z \ll 1$. At what temperatures is this assumption valid?
(d) Find the density of free particles in the low-temperature and high-temperature limits.

**9.7** At room temperature the electron distribution in a metal is close to that at absolute zero. There are relatively few electrons excited above the Fermi energy, leaving holes in the Fermi sea. Show that the probability $P(\Delta)$ of finding an electron with energy $\Delta$ above the Fermi energy is equal to the probability $Q(\Delta)$ of finding a hole at energy $\Delta$ below the Fermi energy.

**9.8** Consider a 2D Fermi gas consisting of $N$ electrons of spin $\frac{1}{2}$, confined in a box of area $A$.

(a) What is the density of single-particle states in momentum space?
(b) Calculate the density $D(\epsilon)$ of single particle states in energy space, and sketch the result.
(c) Find the Fermi energy and Fermi momentum.

(d) Find the internal energy at $T = 0$ as a function of $N$ and $A$.

(e) Find the surface tension at $T = 0$ as a function of $N$ and $A$. (*Hint*: $dE = T\,dS + dA + \mu\,dN$.)

(f) In 3D the chemical potential is proportional to $T^2$ near $T = 0$. Will the behavior in 2D be stronger, weaker, or the same?

# Chapter 10

# The Bose gas

## 10.1 Photons

Photons are the quanta of the electromagnetic field. They are bosons whose number is not conserved, for they may be created and absorbed singly. The Lagrange multiplier corresponding to total number is absent, and the chemical potential $\mu$ is zero. This means that the particles can disappear into the vacuum.

A photon travels at the speed of light $c$, with energy–momentum relation

$$\epsilon = cp, \tag{10.1}$$

where $p$ is the magnitude of the momentum $\mathbf{p}$. The wave number $k$ and the frequency $\omega$ are defined through

$$p = \hbar k = \frac{\hbar \omega}{c}. \tag{10.2}$$

Its intrinsic spin is $\hbar$, but it has only two spin states corresponding to the states of left and right circular polarization. Any enclosure with walls made of matter must contain a gas of photons at the same temperature as the walls, due to emission and absorption of photons by the atoms in the walls. Such a system is called a "black-body cavity", and the photon it contains is called "black-body radiation".

The average occupation number of the state with momentum $\mathbf{p}$ is

$$n_{\mathbf{p}} = \frac{1}{e^{\beta cp} - 1}. \tag{10.3}$$

The absence of the chemical potential makes thermodynamic calculations very simple. The average number of photons is

$$N = 2 \sum_{\mathbf{p}} n_{\mathbf{p}} = 2V \int \frac{d^3 p}{h^3} \frac{1}{e^{\beta cp} - 1}, \tag{10.4}$$

where $V$ is the volume of the system, and the factor 2 comes from the states of polarization. The density of photons is

$$n = \frac{2}{(2\pi)^3} \int_0^\infty dk \frac{4\pi k^2}{e^{\beta \hbar ck} - 1} = \frac{1}{\pi^2 c^3} \int_0^\infty d\omega \frac{\omega^2}{e^{\beta \hbar \omega} - 1} = \kappa \left( \frac{kT}{\hbar c} \right)^3, \tag{10.5}$$

with

$$\kappa = \frac{1}{\pi^2} \int_0^\infty dx \frac{x^2}{e^x - 1} = 4\zeta(3) \approx 0.23, \tag{10.6}$$

where $\zeta(z)$ is the Riemann zeta function.

The internal energy is given by

$$U = 2 \sum_{\mathbf{p}} \epsilon_{\mathbf{p}} n_{\mathbf{p}} = \frac{8\pi c V}{h^3} \int_0^\infty dp \frac{p^3}{e^{\beta cp} - 1}$$

$$= \frac{V\hbar}{\pi^2 c^3} \int_0^\infty d\omega \frac{\omega^3}{e^{\beta\hbar\omega} - 1}, \tag{10.7}$$

which leads to *Stefan's law*

$$\frac{U}{V} = \sigma T^4,$$

$$\sigma = \frac{\pi^2 k^4}{15(\hbar c)^3}, \tag{10.8}$$

where $\sigma$ is called *Stefan's constant*. The specific heat (per unit volume) is therefore

$$c_V = 4\sigma T^3. \tag{10.9}$$

The energy density can be expressed in the form

$$\frac{U}{V} = \int_0^\infty d\omega \, u(\omega, T), \tag{10.10}$$

where $u(\omega, T)$ is the energy density per unit frequency interval:

$$u(\omega, T) = \frac{\hbar}{\pi^2 c^3} \frac{\omega^3}{e^{\beta\hbar\omega} - 1}. \tag{10.11}$$

This is the *Planck distribution*.

The pressure can be obtained through the method described in Section 8.9 modified for photon kinematics. Consider photons reflecting from a wall normal to the $x$-axis. For photons of momentum $\mathbf{p}$, which make an angle $\theta$ with the $x$-axis, the flux of photons is $c \cos \theta$, and the momentum imparted to the wall per reflection is $2p \cos \theta$. Therefore the pressure is given by

$$P = 2 \int_{p_x > 0} \frac{d^3 p}{h^3} 2pcn_{\mathbf{p}} \cos^2\theta, \tag{10.12}$$

where a factor 2 comes from the two polarizations. The restriction $p_x > 0$ means $0 < \theta < \pi/2$. Going to spherical coordinates, we get a factor $\frac{1}{3}$ from the $\theta$

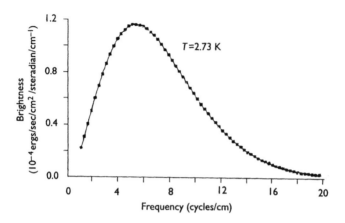

*Figure 10.1* Observed frequency spectrum of cosmic background radiation. Solid line is a Planck distribution with $T = 2.73$ K. (The unit cm is $\frac{1}{3} \times 10^{-10}$ s, the time for light to travel 1 cm.) After Mather (1990).

integration, and obtain

$$P = \frac{8\pi c}{3h^3} \int_0^\infty dp \frac{p^3}{e^{\beta c p} - 1}.$$
(10.13)

Comparison with (10.7) leads to the relation

$$P = \frac{1}{3} \frac{U}{V}.$$
(10.14)

This relation also holds for massive particles at temperatures so high that they move at close to light speed. It is to be compared with the relation $P = \frac{2}{3} U/V$ for a nonrelativistic gas of particles.

Our universe is filled uniformly with black-body radiation, as can be inferred from the measurements shown in Fig. 10.1. The experimental points fit perfectly a Planck distribution of temperature $2.735 \pm 0.05$ K, called the "cosmic background radiation", which is thought to be a relic of the Big Bang.

## 10.2  Bose enhancement

We have seen in Section 8.13 that there is a statistical attraction between free bosons, even though there is no interaction potential. The difference between the Planck distribution and the Maxwell–Boltzmann distribution can be attributed to this attraction, as Einstein showed, in what has come to be known by the historical but clumsy name of "theory of A and B coefficients".

Consider a two-level atom in the wall of a black-body cavity, with level a at a higher energy than level b. Suppose the energy difference is $\hbar\omega$, so that the populations of the two levels in equilibrium are maintained by emission or absorption of

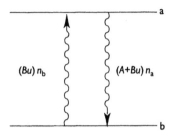

Figure 10.2 Population of atomic levels maintained by photon exchange. The photon distribution function is denoted by u, and A and B are the Einstein coefficients.

photons of frequency $\omega$ in the black-body radiation. "Bose enhancement" refers to the fact that the rates of these processes are enhanced by the presence of photons of the same frequency.

Let $n_a$ and $n_b$ denote, respectively, the number of atoms in levels a and b. Einstein postulates the following rate equations:

$$\frac{dn_a}{dt} = -(A + Bu)\, n_a + Bun_b,$$

$$\frac{dn_b}{dt} = (A + Bu)\, n_a - Bun_b,$$

(10.15)

where $u = u(\omega, T)$ is the energy density of photons per unit frequency. The gain and loss for level a is illustrated in Fig. 10.2. Since atoms leaving a must go to b, and vice versa, we must have $d\,(n_a + n_b)\,/dt = 0$. The coefficient $A$ is the rate of *spontaneous emission*, an intrinsic process independent of the environment. The coefficient $B$ expresses Bose enhancement. It gives the rate of *stimulated emission* induced by the presence of other photons of the same frequency.

In equilibrium we must have $dn_a/dt = dn_b/dt = 0$, which requires

$$(A + Bu)\, n_a = Bun_b.$$

(10.16)

On the other hand, the ratio of the populations should be given by the Boltzmann factor

$$\frac{n_a}{n_b} = e^{-\beta\hbar\omega}.$$

(10.17)

Combining the last two equations, we can solve for $u$, and obtain

$$u(\omega, T) = \frac{A}{B} \frac{1}{e^{\beta\hbar\omega} - 1}.$$

(10.18)

Comparison with (10.11) gives $A/B = \hbar\omega^3/\pi^2 c^3$. If we had omitted the stimulated emission term $Bu$ on the left-hand side of (10.16), we would have obtained the Maxwell–Boltzmann distribution $u = (A/B)\, e^{-\beta\hbar\omega}$.

## 10.3   Phonons

A solid can be idealized as a crystal lattice of atoms, and the small-amplitude vibrations can be described in terms of normal modes, which are harmonic oscillators. The quanta of these normal modes are called *phonons*. They obey Bose statistics, and their number is not conserved. A phonon of momentum $\mathbf{p}$ has energy

$$\epsilon_\mathbf{p} = cp, \tag{10.19}$$

where $p = |\mathbf{p}|$, $c$ is the propagation velocity, and $\omega$ is the frequency. There are three modes of oscillation: one longitudinal mode corresponding to sound waves, and two transverse modes. We have taken the velocities of all three modes to be equal for simplicity.

The difference between a phonon and a photon, apart from the existence of a longitudinal mode, is that the frequency spectrum of a phonon has an upper cutoff, owing to the finite lattice spacing of the underlying crystal. Consider the simple example of beads on a string, as shown in Fig. 10.3. We see that the half-wavelength of oscillations cannot be less than the lattice spacing.

Debye proposes the following model for the normal modes of a solid. Consider waves in a box of volume $V$, with periodic boundary conditions. The modes are

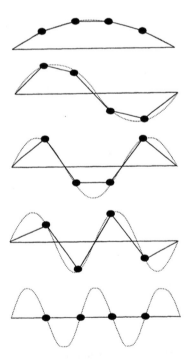

*Figure 10.3* The collective oscillations of a finite number of particles have a minimum wavelength, hence a cutoff frequency.

plane waves labeled by a wavevector $\mathbf{k}$. The number of modes in the element $d^3 k$ is $V d^3 k / (2\pi)^3$. The frequency distribution function $f(\omega)$ is defined by

$$f(\omega)\, d\omega \equiv \frac{3V}{(2\pi)^3} 4\pi k^2 dk, \tag{10.20}$$

where $\omega = ck$ and the factor 3 takes into account the longitudinal and transverse modes. Thus,

$$f(\omega) = \frac{3\omega^2 V}{2\pi^2 c^3}. \tag{10.21}$$

The total number of modes of the system must be equal to $3N$, where $N$ is the number of atoms in the lattice. The cutoff frequency $\omega_m$, or *Debye frequency*, is defined by the requirement

$$\int_0^{\omega_m} f(\omega)\, d\omega = 3N \tag{10.22}$$

which gives

$$\omega_m = \left(6\pi^2 n\right)^{1/3}, \tag{10.23}$$

where $n = N/V$ is the density of atoms. The characteristic energy $\hbar\omega_m$ defines the *Debye temperature* $T_D$:

$$kT_D = \hbar\omega_m. \tag{10.24}$$

Real solids have a more complicated frequency spectrum due to specifics of the crystalline structure, but Debye's simple model reproduces the quadratic behavior at low frequencies, as shown in Fig. 10.4.

In the Debye model, a solid at low temperature is represented by a collection of phonons. The picture becomes inaccurate at higher temperatures, when phonon interactions become important, and breaks down altogether when the solid melts.

## 10.4  Debye specific heat

In the Debye model the internal energy per atom is given by

$$\frac{U}{N} = \frac{1}{N}\sum_i \epsilon_i n_i = \frac{3V}{N(2\pi)^3}\int_0^{k_m} dk(4\pi k^2)\frac{\hbar\omega}{e^{\beta\hbar\omega} - 1}$$

$$= \frac{12\pi\hbar}{n(2\pi)^3}\int_0^{\omega_m} d\omega \frac{\omega^3}{e^{\beta\hbar\omega} - 1}. \tag{10.25}$$

By changing the variable of integration to $t = \beta\hbar\omega$, and introducing the variable

$$u \equiv \frac{T_D}{T}, \tag{10.26}$$

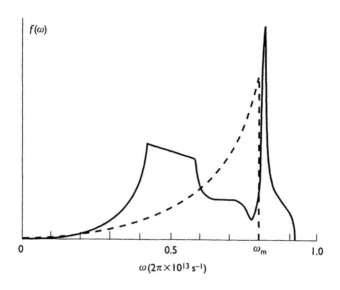

Figure 10.4 The frequency spectrum of Al. The dashed curve is that of the Debye theory, and the solid curve is experimental. After Walker (1956).

we can rewrite the energy per atom in the form

$$\frac{U}{N} = 3kT\, D(u),$$ (10.27)

where $D(u)$ is the Debye function

$$D(u) = \frac{3}{u^3} \int_0^u dt\, \frac{t^3}{e^t - 1}.$$ (10.28)

$D(u)$ has the asymptotic behaviors

$$D(u) \approx 1 - \frac{3}{8}u + \frac{1}{20}u^2 + \cdots \quad (u \ll 1),$$

$$D(u) \approx \frac{\pi^4}{5u^3} + O(e^{-u}) \quad (u \gg 1).$$ (10.29)

The specific heat in the Debye model is given by

$$\frac{C_V}{Nk} = 3D(u) + 3T\frac{dD(u)}{dT}$$ (10.30)

which is a universal function of $T/T_D$. The universal character is verified by experiments, as we can see from the plot in Fig. 10.5. Near absolute zero, the specific heat behaves like $T^3$, which is a reflection of the linear dispersion law

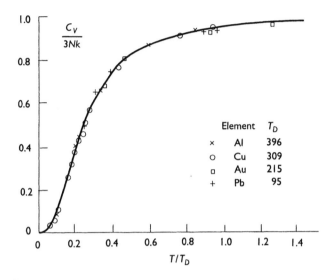

Figure 10.5 Universal curve for the specific heat in the Debye model. Data for different elements are fitted by adjusting the Debye temperature $T_D$. After Wannier (1966).

$\epsilon = cp$ for phonons. It deviates from this form at higher temperatures, when the effect of the cutoff becomes important.

The high-temperature asymptote $3Nk$ conforms to the *law of Dulong and Petit*. This value can be understood in terms of the equipartition of energy among $3N$ harmonic oscillators, each with energy $p^2 + q^2$ (in appropriate units), and thus contributing a term $k$ to the heat capacity. Of course, the high-temperature limit here means that all degrees of freedom of the lattice are fully excited, but not so high that it melts – an effect that lies outside the scope of this model.

## 10.5  Electronic specific heat

There are electrons in the lattice of a solid, and they contribute to the specific heat of the solid. How important the contribution is depends on the temperature, and at room temperature it is completely negligible. As we can see from Fig. 10.5, the Debye specific heat at $T = 300\,\text{K}$ for common metals is of order $3k$, but the electronic contribution is of order $(T/T_F)k$, with $T_F \sim 40,000\,\text{K}$. Thus the electronic contribution amounts to less than 1%. As the temperature decreases, however, the electronic contribution becomes more important. If we expand the specific heat in powers of $T$, then up to order $T^3$ it should have the form

$$\frac{C_V}{Nk} = c_1 \left(\frac{T}{T_F}\right) + c_2 \left(\frac{T}{T_F}\right)^2 + c_3 \left(\frac{T}{T_F}\right)^3 + c_4 \left(\frac{T}{T_D}\right)^3 + \cdots \qquad (10.31)$$

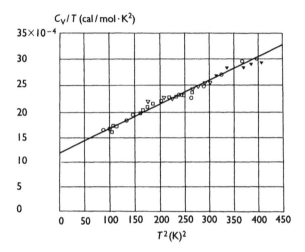

Figure 10.6 Specific heat of KCl. Data for different elements are fitted by adjusting the Debye temperature $T_D$. After Wannier (1966).

where the first three terms come from the electronic Fermi gas, and the last term comes from the Debye model. The $c_n$ are numerical coefficients. Since $T_F \gg T_D$, we can neglect the two terms in the middle, and approximate the above by

$$\frac{C_V}{Nk} \approx aT + bT^3. \tag{10.32}$$

Thus, at low temperatures, a plot of $C_V/T$ against $T^2$ should give a straight line. This is indeed borne out by experiments, as shown in Fig. 10.6. The density of electrons can be obtained from the data. (See Problem 10.10.)

## 10.6   Conservation of particle number

Why are some particle numbers conserved and others not? It depends entirely on their interactions. Phonons in a solid are not "real" particles, for we know that they are collective modes of the atoms in the solid. They are created and absorbed singly, because that is how they interact with the atomic lattice. On the other hand, photons and electrons are considered "real", because we have no evidence to the contrary.

Photons are created and absorbed singly by charged particles. Electrons can be created and destroyed only together with its antiparticle, the positron, by emission of a photon. Thus, the number of electrons minus the number of positrons is conserved. If there are no positrons present, or if the energy of 1 MeV required to create an electron–positron pair is not available, then the electron number is effectively conserved. Similarly, the number of baryons (of which the nucleon is the lightest example) minus the number of antibaryons is conserved. The pattern of fundamental interactions is such that bosons can be created or annihilated singly, while fermions are created or annihilated only in association with an antifermion. (See Huang 1992.)

Atoms are stable in the everyday world, because there are no antinucleons present. It requires at least 2 GeV to create a nucleon–antinucleon pair, and this greatly exceeds the thermal energy at room temperature. Were it not for the conservation of baryons, the bundle of energy locked up in the rest mass of the proton would prefer to assume other forms that give a larger entropy. The hydrogen atom would quickly disappear in a shower of mesons and gamma rays. We can use the Boltzmann factor to estimate the probability for finding a proton of rest mass $M$ at room temperature, if there were no conservation law:

$$e^{-Mc^2/kT} \approx e^{-3\times 10^{10}}. \tag{10.33}$$

This gives us an appreciation of the importance of baryon conservation.

## Problems

**10.1** *Review of quantum harmonic oscillator*: The modes of the electromagnetic field in a cavity, or the phonon modes in a solid, can be described by quantized harmonic oscillators. Consider one such mode with Hamiltonian

$$H = \frac{1}{2m}p^2 + \frac{m\omega^2}{2}q^2,$$

where $p, q$ are hermitian operators defined by the commutation relation $[p, q] = -i\hbar$.

(a) Make the transformation to annihilation operator $a$ and creation operator $a^\dagger$:

$$p = \sqrt{\frac{\hbar m\omega}{2}}\left(a + a^\dagger\right),$$

$$q = i\sqrt{\frac{\hbar}{2m\omega}}\left(a - a^\dagger\right).$$

Show that

$$H = \hbar\omega\left(a^\dagger a + \tfrac{1}{2}\right),$$

$$[a, a^\dagger] = 1.$$

(b) Show that the eigenvalues of $a^\dagger a$ are integers $n = 0, 1, 2, \ldots$.
(c) Let $|n\rangle$ be an eigenstate of $a^\dagger a$ belonging to the eigenvalue $n$. Show that

$$a|n\rangle = \sqrt{n}|n - 1\rangle,$$

$$a^\dagger|n\rangle = \sqrt{n + 1}|n + 1\rangle.$$

The factors $\sqrt{n}, \sqrt{n + 1}$ lead to Bose enhancement.

**10.2** A black body is an idealized object that absorbs all radiation falling upon it, and emits radiation according to Stefan's Law.

A star is maintained by internal nuclear burning at an absolute temperature $T$. It is surrounded by a dust cloud heated by radiation from the star. Treating both the star and the dust cloud as black bodies, show that in radiative equilibrium

(a) the dust cloud reduces the radiative flux from the star to the outside world by half;

(b) the temperature of the dust cloud is $2^{-1/4}T$.

**10.3** The Earth's temperature is maintained through heating by the Sun. Treating both Sun and Earth as black bodies, show that the ratio of the Earth's temperature to the Sun's is given by

$$\frac{T_{\text{Earth}}}{T_{\text{Sun}}} = \sqrt{\frac{R_{\text{Sun}}}{2L}},$$

where $R_{\text{Sun}}$ is the radius of the Sun, and $L$ is the distance between the Sun and the Earth.

**10.4** A window pane in a house has reflection coefficient $r$, i.e. it reflects a fraction $r$, and transmits a fraction $1 - r$ of incident radiation. The temperature outside is $T_0$. Assume there is no heat source in the house, and neglect heat transfer inside the house due to conduction and convection. Show that the temperature just inside the window is

$$T = T_0 \left( \frac{1 - r}{1 - 2r} \right)^{1/4}.$$

**10.5** Show that the entropy of a photon gas is given by $S = \frac{4}{3}U/T$, hence the entropy density is

$$\frac{S}{V} = \frac{4}{3}\sigma T^3,$$

where $\sigma = \pi^2 k^4 / (15(\hbar c)^3)$ is Stefan's constant.

**10.6** The background cosmic radiation has a Planck distribution temperature with temperature 2.73 K, as shown in Fig. 10.1.

(a) What is the photon density in the universe?

(b) What is the entropy per photon?

(c) Suppose the universe expands adiabatically. What would the temperature be when the volume of the universe doubles?

**10.7** *Einstein model of solid*: The Einstein model is a cruder version of the Debye model, in which all phonons have the same frequency $\omega_0$. The simplicity of the model makes it useful for qualitative understanding.

(a) Give the internal energy of this model.

(b) Find the heat capacity. How does it behave near absolute zero?

(c) Show that the free energy is given by

$$A = 3NkT \ln\left(1 - e^{-\beta\hbar\omega_0}\right).$$

**10.8**    A solid exists in equilibrium with its vapor, which is treated as a classical ideal gas. Assume that the solid has a binding energy $\epsilon$ per atom. Use the Einstein model.

(a)  Give the free energy for the total system consisting of solid and vapor. (For the classical ideal gas, you must use correct Boltzmann counting.)
(b)  What is the condition for equilibrium between vapor and solid?
(c)  Find the vapor pressure $P(T)$.

**10.9**    A model of a solid taking into account the structural energy of the lattice postulates a free energy of the form

$$A(V,T) = \phi(V) + A_{\text{phonon}},$$

where $\phi(V)$ is a structural energy, and $A_{\text{phonon}}$ is the contribution from phonons. At absolute zero, there are no phonons, and the volume of the solid is determined by minimizing $\phi(V)$. Near the minimum we may write

$$\phi(V) = -\phi_0 + \frac{K}{2}(V - V_0)^2.$$

At higher temperatures, there is a shift in the minimum due to the fact that the phonon frequency increases when the lattice spacing decreases:

$$\frac{d\omega(k)}{\omega(k)} = -\gamma \frac{dV}{V},$$

where $\gamma$ is known as Gruneisen's constant.

(a)  Neglect the vapor pressure and take the pressure of the solid to be zero. Find the equilibrium condition using the Einstein model for the phonon contribution.
(b)  Find the equilibrium volume of the solid as a function of temperature, and calculate the coefficient of thermal expansion.

**10.10**    In the specific heat formula (10.32) for solids $C_V/Nk = aT + bT^3$, the first term represents the contribution from electrons in the solid, and the second from phonons.

(a)  Give a in terms of the electron density $n$.
(b)  Give b in terms of the Debye temperature $T_D$.
(c)  Obtain the numerical values of a and b from the graph in Fig. 10.6 for KCl, and find $n$ and $T_D$

**10.11**    Surface waves on liquid helium have a dispersion relation given by

$$\epsilon(k) = \hbar \sqrt{\frac{\sigma k^3}{\rho}}$$

where $k$ is the wavenumber of the surface wave, $\sigma$ the surface tension, and $\rho$ the mass density of the liquid. Treating the excitations as bosons with no number conservation, find the internal energy per unit area as a function of temperature.

# Chapter 11

# Bose–Einstein condensation

## 11.1  Macroscopic occupation

Consider a gas of bosons whose total number is conserved. The equation for the fugacity is

$$n\lambda^3 = g_{3/2}(z) \tag{11.1}$$

and a qualitative plot of the function $g_{3/2}$ is shown in Fig.11.1. The function has infinite slope at $z = 1$, since

$$z\frac{d}{dz}g_{3/2}(z) = g_{1/2}(z) = \sum_{\ell=1}^{\infty} \frac{z^\ell}{\sqrt{\ell}} \tag{11.2}$$

and the series diverges at $z = 1$. It is thus bounded by the value

$$g_{3/2}(1) = \sum_{\ell=1}^{\infty} \ell^{-3/2} = \zeta(3/2) = 2.612\ldots \tag{11.3}$$

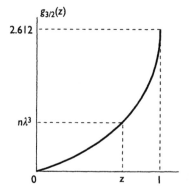

Figure 11.1 This function refers to particles with non-zero momentum, and is bounded by $g_{3/2}(1) = 2.612$. When $n\lambda^3$ exceeds this value, particles must go into the zero-momentum state, leading to Bose–Einstein condensation.

where $\zeta(z)$ is the Riemann zeta function. There is no solution to (11.1) unless

$$n\lambda^3 \le g_{3/2}(1). \tag{11.4}$$

This condition can be violated by increasing the density, or lowering the temperature. What then would be the value of $z$?

To answer this question, recall that in deriving (11.1) in Section 8.12, we have passed to the infinite-volume limit and replaced the sum over states by an integration. This assigns zero weight to the state of zero momentum, for the volume element $4\pi p^2 \, dp$ vanishes at $p = 0$. In the formula (11.1), $n$ is actually the density of particles with non-zero momentum, and this is the quantity bounded by (11.4). When this bound is exceeded, the excess particles cannot disappear, as photons would, because their number is conserved. They are forced to go into the zero-momentum state, and this single quantum state becomes *macroscopically occupied*, i.e. occupied by a finite fraction of the particles. This phenomenon is called *Bose–Einstein condensation*.

The fugacity $z$ is stuck at 1 during the Bose–Einstein condensation. Thus,

$$z = \begin{cases} \sqrt{n\lambda^3} = g_{3/2}(z) & \text{if } n\lambda^3 < g_{3/2}(1) \quad \text{(Gas phase),} \\ 1 & \text{if } n\lambda^3 \ge g_{3/2}(1) \quad \text{(Condensed phase).} \end{cases} \tag{11.5}$$

The critical value $n\lambda^3 = g_{3/2}(1)$ corresponds to

$$n\left(\frac{2\pi\hbar^2}{mkT}\right)^{3/2} = \zeta(3/2). \tag{11.6}$$

This gives the boundary between the gas phase and condensed phase, as shown in the $n$–$T$ diagram of Fig. 11.2. At fixed density $n$, the critical line defines a critical temperature $T_c$:

$$kT_c = \frac{2\pi\hbar^2}{m}\left[\frac{n}{\zeta(3/2)}\right]^{2/3}. \tag{11.7}$$

At a given temperature $T$, it defines a critical density $n_c$:

$$n_c = \zeta(3/2)\left(\frac{mkT}{2\pi\hbar^2}\right)^{3/2}. \tag{11.8}$$

The chemical potential $\mu = kT \ln z$ is zero in the condensed phase. Its qualitative behavior in comparison with the Fermi gas and classical gas is shown in Fig. 11.3.

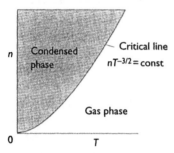

*Figure 11.2* Phase diagram of Bose–Einstein condensation in the density–temperature plane.

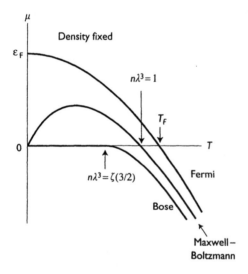

*Figure 11.3* Qualitative behavior of the chemical potential for the Bose, Fer ideal gases.

## 11.2 The condensate

The Bose–Einstein condensate consists of particles in the zero-mo Their number $n_0$ is determined through the relation

$$N = n_0 + \frac{V}{\lambda^3} g_{3/2}(1).$$

Thus,

$$\frac{n_0}{N} = 1 - \frac{g_{3/2}(1)}{n\lambda^3}$$

$$= 1 - \frac{\zeta(3/2)}{n} \left( \frac{mkT}{2\pi\hbar^2} \right)^{3/2}$$

$$= 1 - \left( \frac{T}{T_c} \right)^{3/2}. \tag{11.10}$$

This is illustrated in Fig.11.4.

The momentum distribution $p^2 n_p$ is shown in Fig. 11.5 for both $T > T_c$ and $T < T_c$. The area under the curve gives the density of particles $4\pi h^{-3} \int_0^\infty dp\, p^2 n_p$. Above $T_c$ the graph looks qualitatively like the Maxwell–Boltzmann distribution. Below $T_c$ a $\delta$ function appears at zero momentum, representing the condensate,

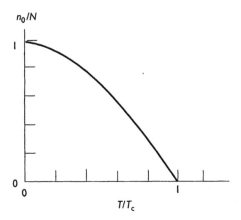

Figure 11.4 Condensate fraction – the fraction of particles with momentum zero.

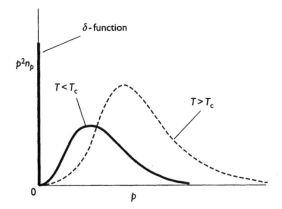

Figure 11.5 Area under the curve is proportional to the density of particles. Below the transition temperature, a delta-function appears at zero momentum, representing the condensate contribution.

and its strength increases as $T \rightarrow 0$, and at absolute zero only the $\delta$ function remains. Because we are plotting $p^2$ times the occupation number, it has the rather singular form $p^{-2}\delta(p)$.

## 11.3   Equation of state

The equation of state as a function of fugacity was derived in Section 8.12:

$$\frac{P}{kT} = \frac{g_{5/2}(z)}{\lambda^3}. \tag{11.11}$$

This is valid in both the gas and condensed phase, because particles with zero-momentum do not contribute to the pressure. Since $z = 1$ in the condensed phase, the pressure becomes independent of density:

$$\frac{P}{kT} = \frac{g_{5/2}(1)}{\lambda^3}, \tag{11.12}$$

where $g_{5/2}(1) = 3.413 \cdots$.

In the gas phase we have to solve for $z$ from (11.1). At low densities, which corresponds to small $z$, the system approaches a classical ideal gas. At higher densities, we have to obtain the solution through numerical methods. The qualitative isotherms are shown in the $P$–$V$ diagram in Fig. 11.6. The Bose–Einstein condensation appears as a first-order phase transition. The vapor pressure is independent of the density:

$$P = C_0 T^{5/2},$$
$$C_0 = g_{5/2}(1)k^{5/2}\left(\frac{m}{2\pi\hbar^2}\right)^{3/2}. \tag{11.13}$$

It can be verified that the vapor pressure satisfies the Clapeyron equation.

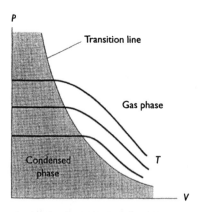

Figure 11.6 Qualitative isotherms of the ideal Bose gas. The Bose–Einstein condensation shows up as a first-order phase transition.

The horizontal part of an isotherm in Fig. 11.6 represents a mixture of the gas phase and the pure condensate, and the latter occupies zero volume because there is no interparticle interaction. In a real Bose gas, which has non-zero compressibility, the Bose–Einstein condensation would be a second-order, instead of a first-order phase transition. This has been observed indirectly in liquid helium, which we shall describe later, and more explicitly in dilute atomic gases trapped in an external potential, which will be discussed in Chapter 15.

## 11.4 Specific heat

The internal energy is given by

$$U = \tfrac{3}{2}PV. \tag{11.14}$$

The heat capacity per particle at constant volume is therefore

$$\frac{C_V}{Nk} = \frac{3}{2nk} \left( \frac{\partial P}{\partial T} \right)_V. \tag{11.15}$$

We shall calculate this separately in the condensed and gas phase. In the condensed phase $P = C_0 T^{5/2}$. Therefore

$$\frac{C_V}{Nk} = \frac{15C_0}{4nk} T^{3/2} \quad (T < T_c). \tag{11.16}$$

The behavior near absolute zero should be compared with other types of behavior we know:

- $T^{3/2}$, for conserved bosons with $\epsilon \propto p^2$
- $T$, for conserved fermions with $\epsilon \propto p^2$
- $T^3$, for non-conserved bosons with $\epsilon \propto p$.

In the gas phase we have

$$
\begin{aligned}
\left( \frac{\partial P}{\partial T} \right)_V &= \left( \frac{\partial}{\partial T} \frac{kT}{\lambda^3} \right)_V g_{5/2}(z) + \frac{kT}{\lambda^3} \left( \frac{\partial}{\partial T} g_{5/2}(z) \right)_V \\
&= \frac{5k}{2} \frac{g_{3/2}(z)}{\lambda^3} + \frac{kT}{\lambda^3} \frac{g_{3/2}(z)}{z} \left( \frac{\partial z}{\partial T} \right)_V,
\end{aligned} \tag{11.17}
$$

where $z$ must be regarded as a function of $T$ and $V$: The derivative $(\partial z/\partial T)_V$ can be obtained by differentiating both sides of (11.1), with the result

$$\left( \frac{\partial z}{\partial T} \right)_V = -\frac{3n\lambda^3}{2T} \frac{z}{g_{1/2}(z)}. \tag{11.18}$$

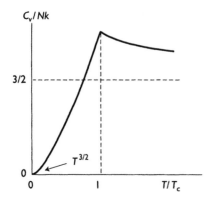

Figure 11.7 Specific heat of ideal Bose gas.

Thus we obtain the result

$$\frac{C_V}{Nk} = \frac{15}{4}\frac{g_{5/2}(z)}{n\lambda^3} - \frac{9}{4}\frac{g_{3/2}(z)}{g_{1/2}(z)} \quad (T > T_c). \tag{11.19}$$

In the high-temperature limit $g_n(z) \approx z \approx n\lambda^3$, and we recover the classical limit $\frac{3}{2}$.

The specific heat is continuous at the transition temperature, as we can verify from (11.16) and (11.19):

$$\frac{C_V}{Nk} \xrightarrow[T \to T_c]{} \frac{15}{4}\frac{g_{5/2}(1)}{g_{3/2}(1)}. \tag{11.20}$$

However, the slope is discontinuous. (See Problem 11.4.) A qualitative plot of the specific heat is shown in Fig. 11.7.

## 11.5  How a phase is formed

The Bose–Einstein condensation of an ideal gas is one of the few examples of a phase transition that can be described completely in terms of microscopic physics. It is interesting to see how the equation of state develops a discontinuous behavior.

To examine the onset of the Bose–Einstein condensation, let us go back to the condition for $z$ at finite volume:

$$\frac{N}{V} = \frac{1}{V}\sum_{\mathbf{p}} \frac{1}{z^{-1}e^{\beta p^2/2m} - 1}. \tag{11.21}$$

In taking the limit $V \to \infty$, we make the replacement

$$\frac{1}{V}\sum_{\mathbf{p}} \to \int \frac{d^3p}{h^3}, \tag{11.22}$$

which assigns zero measure to the state $\mathbf{p} = 0$. In doing so, we have ignored the density of the particles in the zero-momentum state:

$$\frac{n_0}{V} = \frac{1}{V}\frac{z}{1-z}. \tag{11.23}$$

We have a case of non-uniform convergence here, for this quantity approaches different values depending on the order in which we take the limits $V \rightarrow \infty$ and $z \rightarrow 1$:

$$\lim_{z \to 1}\lim_{V \to \infty} \frac{n_0}{V} = 0,$$

$$\lim_{V \to \infty}\lim_{z \to 1} \frac{n_0}{V} = \infty. \tag{11.24}$$

Let us separate out this term in (11.21), and replace the rest by its large $V$ limit:

$$\frac{N}{V} = \frac{1}{V}\frac{z}{1-z} + \int \frac{d^3p}{h^3}\frac{1}{z^{-1}e^{\beta p^2/2m}-1}$$

$$= \frac{1}{V}\frac{z}{1-z} + \frac{1}{\lambda^3}g_{3/2}(z). \tag{11.25}$$

The two terms on the right-hand side are shown in Fig. 11.8, for finite $V$ and for $V \rightarrow \infty$. In the limit $V \rightarrow \infty$, the term $V^{-1}z/(1-z)$ is zero for $z < 1$, and indeterminate at $z = 1$. Thus, the density of zero-momentum particles $n_0/V$ is zero for $z < 1$, but assumes whatever value is required to satisfy (11.21) at $z = 1$.

The fugacity as a function of $n\lambda^3$ is shown in Fig. 11.9. It is a continuous function for finite $V$, however large. In the limit $V \rightarrow \infty$, however, it approaches different functions for $n\lambda^3 > \zeta(3/2)$, and $n\lambda^3 < \zeta(3/2)$. In particular, it is a constant function $z = 1$ in the latter region.

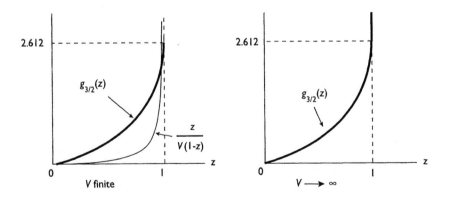

Figure 11.8 In the infinite-volume limit, the condensate occupation number effectively adds a straight vertical rise to $g_{3/2}$ at $z = 1$.

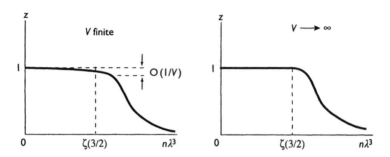

*Figure 11.9* The fugacity approaches different limiting functions for different ranges of $n\lambda^3$, in the infinite-volume limit.

In summary, the equation of state is a regular function of $n$ and $T$ for finite volume, however large. However, in the limit $V \to \infty$ it can approach different limiting forms for different values of $n$ and $T$. Although the idealized discontinuous phase transition occurs only in the thermodynamic limit, macroscopic systems are sufficiently close to that limit as to make the idealization useful.

## 11.6  Liquid helium

Liquid $^4$He exhibits a second order transition, which can be interpreted as a Bose–Einstein condensation modified by strong interatomic interactions. The specific heat has the form shown in Fig. 11.10 and the phase transition is called a "$\lambda$-transition" after the shape of the curve. Below the critical temperature the liquid is a "superfluid", which appears to have zero viscosity, and manifests quantum behavior on a macroscopic scale. The data at the transition point are given by

$$T_c = 2.172 \text{ K}$$

$$n_c = 2.16 \times 10^{22} \text{ cm}^{-3}, \tag{11.26}$$

$$v_c = 46.2 \text{ Å}^3/\text{atom}.$$

Substituting $n_c$ into (11.7) yields $T_c \approx 3.15$ K, which is of the right order of magnitude. Near absolute zero, the specific heat of liquid helium behaves like $T^3$, instead of $T^{3/2}$ as in the ideal Bose gas, because interatomic interactions give rise to phonon excitations.

The low-energy excitations in liquid $^4$He, called "quasiparticles", can be created through neutron scattering. The experimental dispersion relation $\epsilon(k)$, which gives

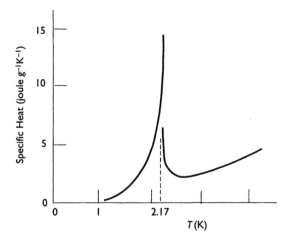

Figure 11.10 Specific heat of liquid ⁴He. The λ-transition at $T = 2.17$ K goes between two liquid phases, both in equilibrium with a gas phase. The specific heat is measured along the vapor pressure curve. After Hill (1957).

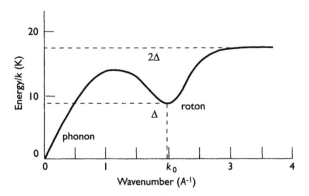

Figure 11.11 Dispersion curve of quasiparticles in liquid ⁴He obtained through neutron scattering. The data points are very dense, with errors about the width of the line plotted. After Donelly (1981) for $0 < k < 2.5$ Å$^{-1}$, Glyde (1998) for $k > 2.5$ Å$^{-1}$.

the energy of the quasiparticle as a function of wave number, is shown in Fig. 11.11. The phonon branch yields

$$\epsilon_{phonon} = \hbar c k,$$
$$c = 239 \pm 5 \text{ m/s}, \tag{11.27}$$

where $c$ is the sound velocity. Near the "roton" minimum, the dispersion curve fits

$$\epsilon_{\text{roton}} = \Delta + \frac{\hbar^2 (k - k_0)^2}{2\sigma};$$

$$\Delta = 8.65 \pm 0.04 \text{ K},$$

$$k_0 = 1.92 \pm 0.01 \text{ A}^{-1},$$  (11.28)

$$\sigma = (0.16 \pm 0.01)\, m,$$

where $m$ is the mass of a helium atom.

## Problems

**11.1**  Verify the Clapeyron equation for $dP/dT$ for the Bose–Einstein condensation.

**11.2**  The Bose functions $g_n(z) = \sum_{\ell=1}^{\infty} \ell^{-n} z^{\ell}$ have the following expansions near $z = 1$:

$$g_{5/2}(z) = 2.363\, v^{3/2} + 1.342 - 2.612\, v - 0.730\, v^2 + \cdots,$$

$$g_{3/2}(z) = 2.612 - 3.455\, v^{1/2} + 1.460\, v + \cdots,$$

where $v = -\ln z$. They are related through $g_{3/2} = -dg_{5/2}/dv$.
  Find the fugacity $z$ of an ideal Bose gas of density $n$, as a function of the temperature $T$ in the neighborhood of $T = T_c$. For $T < T_c$ we have of course $z = 1$. For $T \geq T_c$, obtain the first few terms of a power-series expansion in $T - T_c$.

**11.3**  Show that the equation of state of the ideal Bose gas in the gas phase has the virial expansion

$$\frac{P}{nkT} = 1 + a_2 \left( n\lambda^3 \right) + a_3 \left( n\lambda^3 \right)^2 + \cdots,$$

where

$$a_2 = -\frac{1}{4\sqrt{2}}, \qquad a_3 = \frac{1}{8} - \frac{2}{9\sqrt{3}}.$$

**11.4**  Show that the slope of the heat capacity of an ideal Bose gas has a discontinuity at $T = T_c$ given by

$$\left( \frac{\partial C_V}{\partial T} \right)_{T \to T_c^+} - \left( \frac{\partial C_V}{\partial T} \right)_{T \to T_c^-} = -3.86 \frac{Nk}{T_c}.$$

(*Hint*: Calculate the internal energy via $U = \frac{3}{2} PV$.)

**11.5**    Suppose the particle spectrum has a gap $\Delta > 0$:

$$\epsilon(k) = \begin{cases} -\Delta & (k = 0), \\ \hbar^2 k^2 / 2m & (k > 0). \end{cases}$$

Show how this would modify the Bose–Einstein condensation.

(a) From the equation determining the fugacity, show that the condensation happens when $z = \exp(-\beta\Delta)$.

(b) Assuming $\Delta/kT_0 \ll 1$, where $T_0$ is the transition temperature for $\Delta = 0$, find the shift in transition temperature $T_c - T_0$. Use the expansion of $g_{3/2}(z)$ near $z = 1$, given in Problem 11.2.

**11.6**    The heat capacity of liquid $^4$He near absolute zero should be dominated by contributions from phonons and rotons.

(a) Calculate the contribution $C_{\text{phonon}}$ from a gas of non-conserved bosons, with dispersion relation

$$\epsilon(k) = \hbar ck.$$

(b) Calculate the contribution $C_{\text{roton}}$ from a gas of non-conserved bosons with dispersion relation

$$\epsilon(k) = \Delta + \frac{\hbar^2 (k - k_0)^2}{2\sigma}.$$

**11.7**    Compare $C = C_{\text{phonon}} + C_{\text{roton}}$ to the specific heat data given in the following table. The phonon and roton data from scattering experiments are given in (11.27) and (11.28). The number density and mass density of liquid $^4$He are respectively $n = 2.16 \times 10^{22}$ cm$^{-3}$, and $\rho = 0.144$ g/cm$^3$.

| Temp | Specific heat (J gm$^{-1}$ K$^{-1}$) |
|---|---|
| 0.60 | 0.0051 |
| 0.65 | 0.0068 |
| 0.70 | 0.0098 |
| 0.75 | 0.0146 |
| 0.80 | 0.0222 |
| 0.85 | 0.0343 |
| 0.90 | 0.0510 |
| 0.95 | 0.0743 |
| 1.00 | 0.1042 |

**11.8**    Consider a box divided into two compartments. One compartment contains an almost degenerate ideal Fermi gas of atoms with mass $m_1$ and density $n_1$. The other compartment contains an ideal Bose gas of atoms of mass $m_2$ and density

$n_2$ below the transition temperature for Bose–Einstein condensation. The dividing wall can slide without friction, and conducts heat, so the two gases come to equilibrium with the same pressure and temperature.

(a) Show that the pressure of the Fermi gas has the form

$$P_1 = c_1 \left( \frac{\hbar^2}{m_1} \right) n_1^{5/2},$$

where $c_1$ is a numerical constant.

(b) Show that the vapor pressure of the Bose gas has the form

$$P_2 = c_2 \left( \frac{m_2}{\hbar} \right)^{3/2} (kT)^{5/2},$$

where $c_2$ is a numerical constant.

(c) Give the condition on the temperature for the Fermi gas to be approximately degenerate.

(d) Give the condition for the pressure to equalize.

(e) Find the condition on $m_1/m_2$ in order that (c) and (d) are both satisfied.

**11.9**  By repeating the argument that leads to Bose–Einstein condensation in 3D, show that the transition temperature approaches zero in an ideal Bose gas of particles in 2D.

**11.10**  If the mechanism for photon absorption or emission can be neglected, as may happen in some cosmological settings, the number of photons would be conserved. Can a photon gas undergo Bose–Einstein condensation under these circumstances? If so, give the critical photon density at temperature $T$.

# Chapter 12

# Canonical ensemble

## 12.1 Microcanonical ensemble

The time has come to look beyond the ideal gas, and consider systems with interactions: a dense gas, a liquid, a solid. We already have the principle for doing this, for the microcanonical ensemble defined in Chapter 5 can be applied to any system, not just the ideal gas.

All members of the microcanonical ensemble have the same energy $E$, to within a small tolerance. The entropy of the system is given by

$$S(E) = k \ln \Gamma(E), \tag{12.1}$$

where $k$ is Boltzmann's constant, and $\Gamma(E)$ is the total number of states of the system at energy $E$. The dependence on the number of particles and total volume have been left understood. For thermodynamics, the internal energy $U$ is given by

$$U = E \tag{12.2}$$

and the absolute temperature $T$ by

$$\frac{1}{T} = \frac{\partial S(E)}{\partial E}. \tag{12.3}$$

The formulation holds both for classical and quantum physics.

This method is very convenient when we deal with discrete degrees of freedom, so that calculating $\Gamma(E)$ is a matter of combinatorics, as illustrated by many problems in this chapter. For systems like a real gas, which is specified by a Hamiltonian, the microcanonical ensemble is not as easy to use, and we seek a more convenient ensemble.

## 12.2 Classical canonical ensemble

The key to the new method is to relax the requirement that the energy be fixed, by allowing the system to exchange energy with a heat reservoir. The reservoir and the system form an isolated total system, which can be treated in the microcanonical ensemble, and no new principle is required. What is new is the emphasis. We focus our attention on the subsystem we are interested in, and try to find the probability distribution in its own phase space.

Consider an isolated system divided into two subsystems: a "large" one regarded as a heat reservoir, and a "small" one on which we focus our attention. In the end, of course, even the small system is made to approach the thermodynamic limit. The canonical ensemble is the ensemble for the small system. Let us label the small system 1, and the heat reservoir 2, as illustrated schematically in Fig. 12.1. Neglecting the interactions across the boundary between the two system, we take the total Hamiltonian to be the sum

$$H(p_1, q_1, p_2, q_2) = H_1(p_1, q_1) + H_2(p_2, q_2), \tag{12.4}$$

where $\{p_1, q_1\}$, $\{p_2, q_2\}$, respectively, denote the momenta and coordinates of the two subsystems. The total number of particles $N$, and total energy $E$, are given by

$$N = N_1 + N_2,$$
$$E = E_1 + E_2. \tag{12.5}$$

We keep $N_1$ and $N_2$ separately fixed, but allow $E_1$ and $E_2$ to fluctuate. In other words, the dividing walls between the subsystem allow energy exchange, but not particle exchange. We assume that system 1 is infinitesimally small in comparison with system 2, even though both are macroscopic systems:

$$N_2 \gg N_1,$$
$$E_2 \gg E_1. \tag{12.6}$$

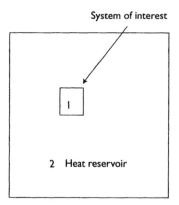

System of interest

1

2   Heat reservoir

Figure 12.1 We focus attention on a "small" system 1, which is part of a larger system. The rest of the system 2 acts as a heat reservoir for system 1.

The phase-space volume occupied by system 2 is given by

$$\Gamma_2(E_2) = \int_{E_2} dp_2\, dq_2, \tag{12.7}$$

where the subscript $E_2$ is shorthand for the condition

$$E_2 < H_2(p_2, q_2) < E_2 + \Delta. \tag{12.8}$$

We are interested in the probability of occurrence of the microstate $\{p_1, q_1\}$ of subsystem 1, regardless of the state of system 2:

$$\text{Probability that 1 is in } dp_1 dq_1 \propto dp_1 dq_1 \Gamma_2(E - E_1). \tag{12.9}$$

This gives the distribution function $\rho_1(p_1, q_1)$ of system 1 in its own $\Gamma$ space:

$$\rho_1(p_1, q_1) = \Gamma_2(E - E_1)$$
$$= \Gamma_2\left(E - H_1(p_1, q_1)\right). \tag{12.10}$$

Since $E_1 \ll E$, we expand the above in powers of $E_1$ to lowest order. It is convenient to expand the logarithm of $\Gamma_2$, which can be expressed in terms of the entropy of system 2:

$$k \ln \Gamma_2(E - E_1) = S_2(E - E_1)$$
$$= S_2(E) - E_1 \left.\frac{\partial S_2(E')}{\partial E'}\right|_{E'=E} + \cdots$$
$$\approx S_2(E) - \frac{E_1}{T}, \tag{12.11}$$

where $T$ is the temperature of system 2. This relation becomes exact in the limit when system 2 becomes a heat reservoir. The temperature of the heat reservoir $T$ fixes the temperature of system 1.

The density function for system 1 is therefore

$$\rho_1(p_1, q_1) = e^{S_2(E)/k} e^{-E_1/kT}. \tag{12.12}$$

The first factor is a constant, which can be dropped by redefining the normalization. In the second factor the energy of the system can be replaced by its Hamiltonian. Thus

$$\rho_1(p_1, q_1) = e^{-\beta H_1(p_1, q_1)}, \tag{12.13}$$

with

$$\beta = \frac{1}{kT}. \tag{12.14}$$

This is the *canonical ensemble*, appropriate for a system with a fixed number of particles, in contact with a heat reservoir at temperature $T$. Since we shall refer

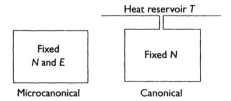

Figure 12.2 Schematic representations of the microcanonical ensemble and canonical ensemble.

only to the small system from now on, the subscript 1 is unnecessary, and will be omitted. Schematic representations of the microcanonical and canonical ensembles are given in Fig. 12.2.

## 12.3   The partition function

The *partition function* for the canonical ensemble is defined by

$$Q_N(T, V) \equiv \int \frac{dp\,dq}{\tau}\, e^{-\beta H(p,q)},$$

$$\tau = N!h^{3N}, \tag{12.15}$$

where $dp = d^{3N}p$, $dq = d^{3N}q$, and each $d^3q$ integration ranges over the volume $V$ of the system. The factor $\tau$ serves two purposes. It makes $Q_N$ dimensionless by supplying the unit $h$ for an elementary cell in phase space, and it supplies a factor $N!$, known as "correct Boltzmann counting", to take into account the indistinguishability of the particle in quantum mechanics.

As we shall see, all thermodynamic information can be obtained from the partition function. In particular, the entropy is completely determined, with no arbitrary constants, and the factor $N!$ will make it an extensive quantity.

## 12.4   Connection with thermodynamics

The thermodynamic internal energy is the ensemble average of the energy:

$$U = \langle H \rangle = \frac{\int dp\,dq\, H\, e^{-\beta H(p,q)}}{\int dp\,dq\, e^{-\beta H(p,q)}}. \tag{12.16}$$

This can be obtained from the partition function by differentiation with respect to $\beta$:

$$U = -\frac{\partial}{\partial \beta} \ln Q_N. \tag{12.17}$$

The general connection between the canonical ensemble and thermodynamics is given by the statement

$$Q_N(T, V) = e^{-\beta A(V,T)}, \tag{12.18}$$

where $A(V, T)$ is the Helmholtz energy.

To show that (12.18) correctly identifies the free energy, let us rewrite it in the form

$$\int \frac{dp\, dq}{\tau} e^{-\beta H(p,q)} = e^{-\beta A(V,T)},$$

$$\int \frac{dp\, dq}{\tau} e^{\beta[A(V,T)-H(p,q)]} = 1. \tag{12.19}$$

Differentiating both sides with respect to $\beta$, we obtain

$$\int \frac{dp\, dq}{\tau} e^{\beta[A(V,T)-H(p,q)]} \left[ A(V,T) + \beta \frac{\partial A(V,T)}{\partial \beta} - H(p,q) \right] = 0. \tag{12.20}$$

The first two terms are not functions of $p, q$ and can be taken outside of the integral, which is equal to 1. The second term gives the internal energy. Thus

$$A(V,T) + \beta \frac{\partial A(V,T)}{\partial \beta} - U = 0,$$

$$A(V,T) - T \frac{\partial A(V,T)}{\partial T} - U = 0. \tag{12.21}$$

This is consistent with the thermodynamic relations

$$S = -\frac{\partial A(V,T)}{\partial T},$$

$$A = U - TS. \tag{12.22}$$

Therefore, we have a correct definition of $A(V,T)$, from which we can obtain all thermodynamic functions through the Maxwell relations.

## 12.5  Energy fluctuations

The systems in the canonical ensemble have different energies, whose values fluctuate about the mean energy $U$ determined by the temperature. To calculate the mean-square fluctuation of the energy we start with the expression

$$U = -\frac{\partial}{\partial \beta} \ln \int dp\, dq\, e^{-\beta H},$$

(12.23)

and differentiate with respect to $\beta$:

$$\frac{\partial U}{\partial \beta} = -\frac{\int dp\, dq H^2\, e^{-\beta H}}{\int dp\, dq\, e^{-\beta H}} + \frac{\left(\int dp\, dq H\, e^{-\beta H}\right)^2}{\left(\int dp\, dq\, e^{-\beta H}\right)^2}$$

$$= -\langle H^2 \rangle + \langle H \rangle^2.$$

(12.24)

Using thermodynamic definitions, we can rewrite

$$\frac{\partial U}{\partial \beta} = \frac{\partial U}{\partial T}\frac{\partial T}{\partial \beta} = -kT^2\frac{\partial U}{\partial T} = -kT^2 C_V.$$

(12.25)

Thus

$$\langle H^2 \rangle - \langle H \rangle^2 = kT^2 C_V.$$

(12.26)

For macroscopic systems the left-hand side is of order $N^2$, while the right-hand side is of order $N$. Energy fluctuations are therefore "normal", and become negligible when $N \to \infty$. This is why the results of the canonical and microcanonical ensembles coincide in that limit.

## 12.6   Minimization of free energy

We have learned in thermodynamics that a system at fixed $V, T$ will seek the state of minimum free energy. This principle can be derived using the canonical ensemble.

Note that the partition function (12.15) is an integral over states, but the integrand depends only on the Hamiltonian, whose value is the energy. We can convert the integral into one over the energy by writing

$$Q_N(V, T) = \int dE\, e^{-\beta E} \int \frac{dp\, dq}{\tau} \delta\left(H(p,q) - E\right).$$

(12.27)

This is a trivial rewriting, for the $dE$ integration can be done by simply setting $E = H(p,q)$. The $dp\, dq$ integral now gives the density of states at energy $E$:

$$\omega(E) = \int \frac{dp\, dq}{\tau} \delta\left(H(p,q) - E\right),$$

(12.28)

which is related to the entropy by

$$S(E) = k \ln \omega(E).$$

(12.29)

Thus

$$Q_N(V,T) = \int dE\, \omega(E)\, e^{-\beta E} = \int dE\, e^{-\beta[E-TS(E)]} \qquad (12.30)$$

or

$$Q_N(V,T) = \int dE\, e^{-\beta A(E)}, \qquad (12.31)$$

where $A(E) \equiv E - TS(E)$.

The dominant contribution to the integral will come from the value of $E$ for which $A(E)$ is minimum. This can be seen as follows. Let the minimum of $A(E)$ occur at $\bar{E}$, which is determined by the condition

$$\left.\frac{\partial A}{\partial E}\right|_{E=\bar{E}} = \left[1 - T\frac{\partial S}{\partial E}\right]_{E=\bar{E}} = 0. \qquad (12.32)$$

This gives

$$\left.\frac{\partial S}{\partial E}\right|_{E=\bar{E}} = \frac{1}{T}. \qquad (12.33)$$

which is just the thermodynamic relation between entropy and temperature. The second derivative of $A(E)$ gives

$$\frac{\partial^2 A}{\partial E^2} = -T\frac{\partial^2 S}{\partial E^2}. \qquad (12.34)$$

From the Maxwell relation $\partial E/\partial S = -T$, we have $\partial S/\partial E = -1/T$. Hence,

$$\frac{\partial^2 S}{\partial E^2} = \frac{1}{T^2}\frac{\partial T}{\partial E} = \frac{1}{T^2 C_V}. \qquad (12.35)$$

Thus, the second derivative is positive:

$$\frac{\partial^2 A}{\partial E^2} = \frac{1}{TC_V} \qquad (12.36)$$

showing that the free energy at $\bar{E}$ has a minimum (and not maximum). Now expand $A(E)$ about the minimum:

$$A(E) = A(\bar{E}) + \frac{1}{2TC_V}\left(E - \bar{E}\right)^2 + \cdots \qquad (12.37)$$

Neglecting the higher-order terms, we have

$$Q_N(V,T) = e^{-\beta A(\bar{E})} \int dE\, e^{-(E-\bar{E})^2/(2kT^2 C_V)}. \qquad (12.38)$$

In the thermodynamic limit $C_V$ becomes infinite, and the integrand is very sharply peaked at $E = \bar{E}$, as illustrated in Fig. 12.3. This is why we can neglect the higher-order terms, and evaluate the integral by extending the limits of integration from

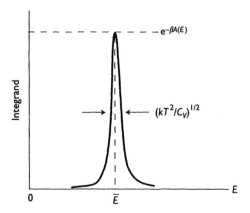

Figure 12.3 When the partition function is expressed as an integral over energy, the integrand is sharply peaked at $E = \bar{E}$ corresponding to a minimum in the free energy $A = E - TS$.

$-\infty$ to $\infty$, to obtain

$$Q_N = \sqrt{2\pi kT^2 C_V}\, e^{-\beta A(\bar{E})},$$

$$\ln Q_N(V, T) = -\frac{A(\bar{E})}{kT} + \frac{1}{2}\ln\left(2\pi kT^2 C_V\right). \tag{12.39}$$

In the thermodynamic limit, the first term is of order $N$, while the second term is of order $\ln N$, and can be neglected. Thus, we see that the system seeks a state of minimum free energy.

## 12.7  Classical ideal gas

We illustrate the classical canonical ensemble with the ideal gas. The Hamiltonian is given by

$$H = \sum_{i=1}^{N} \frac{\mathbf{p}_i^2}{2m}. \tag{12.40}$$

The partition function is

$$Q_N(T, V) = \int \frac{d^{3N}p\, d^{3N}q}{N! h^{3N}} e^{-\beta(\mathbf{p}_1^2 + \cdots + \mathbf{p}_N^2)/2m} \tag{12.41}$$

$$= \frac{V^N}{N!}\left(\int_{-\infty}^{\infty} \frac{dp}{h} e^{-\beta p^2/2m}\right)^{3N} = \frac{1}{N!}\left(\frac{V}{\lambda^3}\right)^N, \tag{12.42}$$

where $\lambda = \sqrt{2\pi\hbar^2/mkT}$ is the thermal wavelength. Thus, the free energy is given by

$$A(V,T) = -kT \ln Q_N(T,V)$$
$$= NkT \left[ \ln(n\lambda^3) - 1 \right], \tag{12.43}$$

where $n = N/V$, and we have used the Stirling approximation $\ln N! \approx N \ln N - N$. The chemical potential is given by the Maxwell relation

$$\mu = \left( \frac{\partial A}{\partial N} \right)_{V,T} = kT \ln \left( n\lambda^3 \right) \tag{12.44}$$

and the entropy is given by

$$S = - \left( \frac{\partial A}{\partial T} \right)_V = Nk \left[ \frac{5}{2} - \ln \left( n\lambda^3 \right) \right]. \tag{12.45}$$

More explicitly,

$$\frac{S}{Nk} = \frac{5}{2} + \ln \frac{V}{N} + \frac{3}{2} \ln \left( \frac{mkT}{2\pi\hbar^2} \right). \tag{12.46}$$

This is the *Sacher–Tetrode equation*, which fixes the defects in the thermodynamic result in Section 2.8.

## 12.8  Quantum ensemble

A state of a quantum-mechanical system is described by a complex wave function, which has a phase, for which there is no classical analog. In the definition of a quantum ensemble, the relative phases of the wave functions are assumed to be random numbers. In practice, this means that there is no quantum interference between two different members of the ensemble.

Consider a quantum system whose instantaneous wave function at time $t$ is $\Psi(x,t;X)$, where $x$ denotes all the coordinates of the system, and $X$ those of the environment, such as a heat reservoir. We can expand the wave function in terms of the eigenfunctions $\psi_n$ of the Hamiltonian $H$ of the system, with energy eigenvalue $E_n$:

$$H\psi_n = E_n\psi_n. \tag{12.47}$$

The expansion has the form

$$\Psi(x,t;X) = \sum_n c_n(X,t)\psi_n(x),$$
$$\tag{12.48}$$
$$\sum_n \int dX \, c_n^* c_n = 1.$$

The expansion coefficients $c_n(X,t)$ contain all the dependences on $t$ and $X$. If the $c_n$ are definite complex numbers, then $\Psi$ corresponds to one pure quantum state.

When the relative phases of $c_n$ are regarded as random numbers, then $\Psi$ represents an incoherent superposition of states – a quantum ensemble.

The phase of $c_n(X, t)$ is of course a definite number. But, due to the complexity of the environment, it is an extremely sensitive function. It changes by a large multiple of $2\pi$, under even the slightest changes in $X$ and $t$. This is the physical reason why the phase is essentially random. This idea is implemented in the following manner.

The expectation values of any observable $O$ is given by

$$(\Psi, O\Psi) = \sum_{n,m} c_n^* c_m O_{nm}, \tag{12.49}$$

where $O_{nm}$ is the matrix element:

$$O_{nm} = \int dx \psi_n^*(x) O \psi_m(x). \tag{12.50}$$

The thermal average is obtained by integrating this over the coordinates of the environment, and averaging the result over time:

$$\langle O \rangle = \overline{\int dX (\Psi, O\Psi)} = \sum_{n,m} O_{nm} \overline{\int dX \, c_n^* c_m}, \tag{12.51}$$

where an overhead bar denotes averaging over a time interval which is large compared with the relaxation time, but small on a macroscopic scale.

The *assumption of random phase* is the statement

$$\overline{\int dX \, c_n^*(X, t) c_m(X, t)} = 0 \quad \text{if } n \neq m. \tag{12.52}$$

That is, the relative phase varies so rapidly and erratically with time that it averages to zero. The thermal average now takes the form

$$\langle O \rangle = \frac{\sum_n \rho_n O_{nn}}{\sum_n \rho_n}, \tag{12.53}$$

where $\rho_n = \overline{\int dX |c_n|^2}$.

The quantum ensemble is defined by $\rho_n$, the relative probabilities for the occurrence of the $n$th energy eigenstate. We can define an operator $\rho$, the *density matrix*, which is diagonal with respect to energy eigenstates, with eigenvalue $\rho_n$:

$$\langle n|\rho|m \rangle = \delta_{nm} \rho_n, \tag{12.54}$$

where $|n\rangle$ is an energy eigenstate of the system. In terms of some other sets of states, of course, $\langle n|\rho|m \rangle$ will not be diagonal, in general. The thermal average can now be written as

$$\langle O \rangle = \frac{\text{Tr}(\rho O)}{\text{Tr}\,\rho}, \tag{12.55}$$

where Tr denotes the sum of diagonal elements. This formula is independent of the matrix representation. That is, it has the same value in any basis, including ones in which $\rho$ is not diagonal.

## 12.9  Quantum partition function

The canonical ensemble corresponds to the choice

$$\rho_n = e^{-\beta E_n} \tag{12.56}$$

The density matrix is given by

$$\rho = \sum_n e^{-\beta E_n} |n\rangle \langle n| = e^{-\beta H} \sum_n |n\rangle \langle n| = e^{-\beta H}, \tag{12.57}$$

where we have used the fact that $\sum_n |n\rangle \langle n| = 1$ for a complete set. The thermal average now takes the form

$$\langle O \rangle = \frac{\text{Tr}\left(O e^{-\beta H}\right)}{\text{Tr}\, e^{-\beta H}}. \tag{12.58}$$

The quantum partition function can be written in the form

$$Q_N(V, T) = \text{Tr}\, e^{-\beta H}, \tag{12.59}$$

where the trace extends over all states whose wave functions satisfy specified boundary conditions, with the correct symmetry as dictated by the statistics of the particles. The number of particles $N$ enters through the Hamiltonian operator $H$, the volume $V$ enters through the boundary condition imposed on the wave functions of the system, and of course the temperature enters through $\beta = 1/kT$. The internal energy is given by

$$U = \langle H \rangle = -\frac{\partial}{\partial \beta} \ln Q \tag{12.60}$$

and the free energy $A(V, T)$ can be obtained through the formula

$$e^{-\beta A} = \text{Tr}\, e^{-\beta H}. \tag{12.61}$$

## 12.10   Choice of representation

The quantum partition function is independent of the choice of representation. That is, we can write

$$\mathrm{Tr}\, e^{-\beta H} = \sum_{\alpha} \langle \alpha | e^{-\beta H} | \alpha \rangle \tag{12.62}$$

where $|\alpha\rangle$ is a member of any complete orthonormal set of states, as long as it satisfies the boundary conditions and symmetries of the system:

$$\langle \alpha | \beta \rangle = \delta_{\alpha\beta},$$

$$\sum_{\alpha} |\alpha\rangle\langle\alpha| = 1. \tag{12.63}$$

If we choose $|\alpha\rangle = |n\rangle$, then since $e^{-\beta H}|n\rangle = e^{-\beta E_n}|n\rangle$, we have

$$\mathrm{Tr}\, e^{-\beta H} = \sum_{n} e^{-\beta E_n}. \tag{12.64}$$

From a formal point of view this appears to be a natural choice. In practical applications, however, the eigenvalues of $H$ are usually unknown. It would then be more practical to choose a set $|\alpha\rangle$ convenient for making approximations.

## Problems

**12.1**   A perfect crystal has $N$ lattice sites and $M$ interstitial locations. An energy $\Delta$ is required to remove an atom from a site and place it in an interstitial, when the number of displaced atoms, $n$, is much smaller than $N$ or $M$.

(a)   How many ways are there of removing $n$ atoms from $N$ sites?
(b)   How many ways are there of placing $n$ atoms on $M$ interstitials?
(c)   Use the microcanonical ensemble to calculate the entropy as a function of total energy $E$, and define the temperature.
(d)   Show that the average number of displaced atoms $n$ at temperature $T$ is given by

$$\frac{n^2}{(N-n)(M-n)} = e^{-\Delta/kT}.$$

Obtain $n$ for $\Delta \gg kT$, and $\Delta \ll kT$.

(e)   Use this model for defects in a solid. Set $N = M$, and $\Delta = 1\,\mathrm{eV}$. find the defect concentration at $T = 1,000$ and $300\,\mathrm{K}$.

**12.2**   A one-dimensional chain, fixed at one end, is made of $N$ identical elements each of length $a$. The angle between successive elements can be either $0°$ or $180°$, as shown in the accompanying sketch. There is no difference in the energies of these two possibilities. We can think of each element as either pointing right ($+$)

or left ( − ). Suppose $N_\pm$ is the number of $\pm$ elements, and $L$ is the total length of the chain. We have

$$N = N_+ + N_-,$$
$$L = a\,(N_+ - N_-).$$

An interesting feature of this model is that the internal energy does not depend on $L$, but there is a tension $\tau$ defined through

$$dU = T\,dS + \tau\,dL.$$

It arises statistically, through the fact that the end wants to be free to sample phase space.

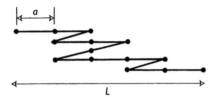

Find the following quantities, using the microcanonical ensemble:
(a)  the entropy as a function of $N$ and $N_+$
(b)  the free energy as a function of $N$ and $N_+$
(c)  the tension $\tau$ as a function of $T, N, L$.

**12.3**   A chain made of $N$ segments of equal length $a$ hangs from the ceiling. A mass $m$ is attached to the other end under gravity. Each segment can be in either of two states, up or down, as illustrated in the sketch.

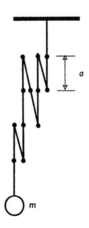

(a) Show that the partition function at temperature $T$ is given by

$$Q_N = \left(1 + e^{-mga/kT}\right)^N.$$

(b) Find the entropy of the chain.

(c) Find the internal energy, and determine the length of the chain.

(d) Show that the chain obeys Hooke's law, namely, a small force pulling on the chain increases its length proportionately. Find the proportionality constant.

**12.4**   The unwinding of a double-stranded DNA molecule is like unzipping a zipper. The DNA has $N$ links, each of which can be in one of two states: a closed state with energy 0, and an open state with energy $\Delta$. A link can open only if all the links to its left are already open, as illustrated in the sketch.

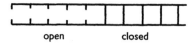

open          closed

(a) Show that the partition function of the DNA chain is

$$Q_N = \frac{1 - e^{-(N+1)\Delta/kT}}{1 - e^{-\Delta/kT}}.$$

(b) Find the average number of open links in the low-temperature limit $kT \ll \Delta$.

**12.5**   Consider a piece of two-dimensional graphite where $N$ carbon atoms form a honeycomb lattice. Assume that it costs energy $\Delta$ to remove a carbon atom from a lattice site and place it in the center of a hexagon to form a vacancy and an interstitial, as shown in the sketch.

(a) Show that there are $N/2$ possible locations for interstitials.

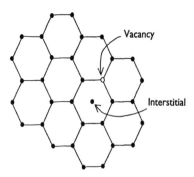

Vacancy

Interstitial

(b) Consider a microcanonical ensemble of the system, at given total energy $E$. For $M$ interstitials, find the statistical entropy for large $N$ and $M$.

(c) Define the temperature by $T^{-1} = \partial S/\partial E$, and find $E$ as a function of $T$, for $T \to 0$ and $T \to \infty$.

**12.6** A particle can exist in only three states labeled $n = 1, 2, 3$. The energies $\epsilon_n$ of these states depend on a parameter $x \geq 0$, with two of the energies degenerate:

$$\epsilon_1 = \epsilon_2 = bx^2 - \tfrac{1}{2}cx,$$

$$\epsilon_3 = bx^2 + cx,$$

where $b$ and $c$ are constants.

(a) Find the Helmholtz free energy per particle $a(x, T) = A_N(x, T)/N$ for a collection of $N$ such particles, assuming that there are no inter-particle interactions.

(b) If $x$ is allowed to freely vary at constant $T$, it will assume an equilibrium value $\bar{x}$ that minimizes the free energy. Find $\bar{x}$ as a function of $T$. Show that there is a phase transition, and find the transition temperature. Assume that $\bar{x}$ is small, and expand $\partial a(x, T)/\partial x$ in a power series in $x$ to order $x^2$.

This model can be used to describe ions in a crystal subject to a uniform strain characterized by the parameter $x$. The phase transition is known as the "cooperative Jahn-Teller" phase transition.

**12.7** Set up the partition function of a classical relativistic ideal gas and obtain the free energy in terms of an integral. In the non-relativistic and the ultra-relativistic limits, show that the chemical potentials are given by

Non-relativistic: $\quad \mu \approx mc^2 + kT \ln(n\lambda^3) \quad (\lambda = \sqrt{2\pi\hbar^2/mkT})$,

Ultra-relativistic: $\quad \mu \approx kT \ln(nL^3) \quad (L = \pi^{2/3}\hbar c/kT)$.

**12.8** Consider a non-relativistic free quantum gas obeying "Boltzmann statistics." This means that the particles are regarded as distinguishable; no particular symmetry is imposed on the wave function with respect to the exchange of particles. (It would be more accurate to refer to this case as "no statistics".) Calculate the quantum partition function, and obtain the equation of state.

**12.9** Consider a collection of $N$ non-interacting one-dimensional harmonic oscillators, with total Hamiltonian

$$H(p, q) = \sum_{i=1}^{N} \left[ \frac{p_i^2}{2m} + \frac{1}{2}m\omega^2 q_i^2 \right].$$

(a) Calculate the classical partition function, taking the phase-space element to be $dp\,dq/\tau$, where $\tau$ is an arbitrary scale factor.

(b) Obtain the entropy, internal energy, and heat capacity.

**12.10**   Consider the $N$ non-interacting harmonic oscillators in quantum statistical mechanics.

(a)  The problem of distinguishable harmonic oscillators is the Einstein model of a solid studied in Problem 10.1. Find the free energy via the partition function. Go to the high-temperature limit, compare the results with that of the last problem, and determine the scale factor $\tau$ in the classical problem.

(b)  Write down the quantum partition function for $N$ indistinguishable harmonic oscillators, for Bose and Fermi statistics. You need not evaluate them.

**12.11**   Consider a system of $N$ non-interacting spins, whose energies in a magnetic field $B$ are given by $\pm\mu_0 B$. Ignore translational motion.

(a)  Calculate the partition function $Q_N$.

(b)  Calculate the average magnetic moment $\langle M \rangle$.

(c)  Find the mean-square fluctuation $\langle M^2 \rangle - \langle M \rangle^2$.

# Chapter 13

# Grand canonical ensemble

## 13.1 The particle reservoir

The grand canonical ensemble is built upon the canonical ensemble by relaxing the restriction to a definite number of particles. The relative probability of finding the system with $N$ particles at temperature $T$, in a volume $V$, is taken to be

$$\rho(N, V, T) = z^N Q_N(V, T), \tag{13.1}$$

where $z = e^{\beta\mu}$ is the fugacity. The chemical potential $\mu$ is a given external parameter, in addition to the temperature. The system exchanges energy with a heat reservoir at temperature $T$, which determines the average energy, and it exchanges particles with a "particle reservoir" of chemical potential $\mu$, which determines the average number of particles. A schematic representation of the ensemble is shown in Fig. 13.1.

The grand canonical ensemble is a more realistic representation of physical systems than the canonical ensemble, for we can rarely fix the total number of particles in a macroscopic system. A typical example is a volume of air in the atmosphere. The particle reservoir in this case is the rest of the atmosphere. The number of air molecules in the volume considered fluctuates about a mean value determined by the rest of the atmosphere.

## 13.2 Grand partition function

The ensemble average in the grand canonical ensemble is obtained by averaging the canonical average over the number of particles $N$. For example, the internal energy is given by

$$U = \frac{\sum E_N z^N Q_N}{\sum z^N Q_N}, \tag{13.2}$$

where $E_N$ is the average energy in the canonical ensemble:

$$E_N = -\frac{\partial}{\partial\beta} \ln Q_N. \tag{13.3}$$

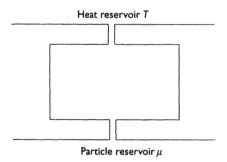

Heat reservoir $T$

Particle reservoir $\mu$

*Figure 13.1* Schematic representation of the grand canonical ensemble. The system exchanges particles with a particle reservoir with fixed chemical potential, and exchanges energy with a heat reservoir at fixed temperature.

It is useful to introduce the *grand partition function*:

$$Q(z, V, T) = \sum_{N=0}^{\infty} z^N Q_N(V, t) \tag{13.4}$$

in terms of which we can write

$$U = -\frac{\partial}{\partial \beta} \ln Q(z, V, T). \tag{13.5}$$

In the thermodynamic limit $V \to \infty$, we expect that

$$\frac{1}{V} \ln Q(z, V, T) \xrightarrow[V \to \infty]{} \text{Finite limit.} \tag{13.6}$$

## 13.3  Number fluctuations

The average number of particles is given by

$$\bar{N} = \frac{\sum N z^N Q_N}{\sum z^N Q_N} = z \frac{\partial}{\partial z} \ln Q(z, V, T). \tag{13.7}$$

Specifying $\bar{N}$ determines the chemical potential $\mu = kT \ln z$. The mean-square fluctuation can be obtained by differentiating again with respect to $z$:

$$z \frac{\partial}{\partial z} z \frac{\partial}{\partial z} \ln Q(z, V, T) = \frac{\sum N^2 z^N Q_N}{\sum z^N Q_N} - \left[ \frac{\sum N z^N Q_N}{\sum z^N Q_N} \right]^2$$

$$= \overline{N^2} - \overline{N}^2.$$

In terms of the chemical potential we can write

$$z\frac{\partial}{\partial z} = z\frac{\partial\mu}{\partial z}\frac{\partial}{\partial\mu} = kT\frac{\partial}{\partial\mu}. \tag{13.8}$$

Thus

$$\overline{N^2} - \overline{N}^2 = (kT)^2 \frac{\partial^2}{\partial\mu^2} \ln\mathcal{Q}(z,V,T). \tag{13.9}$$

Dividing both sides by $V^2$, we have the density fluctuation

$$\overline{n^2} - \bar{n}^2 = \frac{(kT)^2}{V^2}\frac{\partial^2}{\partial\mu^2} \ln\mathcal{Q}(z,V,T). \tag{13.10}$$

Assuming (13.6), we see that this vanishes like $V^{-1}$ in the thermodynamic limit, and makes the grand canonical ensemble equivalent to the canonical ensemble.

## 13.4  Connection with thermodynamics

Assuming that the number fluctuation is vanishingly small, we need to keep only the largest term in the sum over $N$:

$$\ln\mathcal{Q}(z,V,T) = \ln\sum_{N=0}^{\infty} z^N Q_N(V,T) \approx \ln\left[z^{\bar{N}}Q_{\bar{N}}(V,T)\right]$$

$$= \bar{N}\ln z + \ln Q_{\bar{N}}(V,T) = \frac{\bar{N}\mu}{kT} + \ln Q_{\bar{N}}(V,T), \tag{13.11}$$

where $\bar{N}$ is the average number of particles. Now put

$$Q_N(V,T) = e^{-\beta A_N(V,T)},$$
$$A_N(V,T) = Na(v,T),$$

where $a(v,T)$ is the free energy per particle, and $v = V/N$. Then we have

$$\ln\mathcal{Q}(z,V,T) = \frac{\bar{N}}{kT}[\mu - a(\bar{v},T)]; \tag{13.12}$$

where $\bar{v} = V/\bar{N}$.

The pressure is given by a Maxwell relation:

$$P = -\left[\frac{\partial A_N(V,T)}{\partial V}\right]_{N,T} = -\frac{\partial}{\partial V}[Na(v,T)]_{N,T}$$

$$= -N\frac{\partial a(v,T)}{\partial V} = -\frac{\partial a(v,T)}{\partial v}. \tag{13.13}$$

The chemical potential is given by

$$\mu = \left[\frac{\partial A_N(V,T)}{\partial N}\right]_{V,T} = a(v,T) + N\frac{\partial a(v,T)}{\partial N}. \tag{13.14}$$

Since

$$\frac{\partial a(v,T)}{\partial N} = \frac{\partial a}{\partial v}\frac{\partial v}{\partial N} = -\frac{V}{N^2}\frac{\partial a}{\partial v}, \tag{13.15}$$

we have

$$\mu = a(v,T) - v\frac{\partial a(v,T)}{\partial v}. \tag{13.16}$$

Thus we obtain the thermodynamic relation

$$\mu = a(v,T) + Pv. \tag{13.17}$$

Using (13.12) we obtain

$$\ln \mathcal{Q}(z,V,T) = \frac{PV}{kT}, \tag{13.18}$$

and combining with (13.7) leads to the equation of state in parametric form:

$$\frac{P}{kT} = \frac{1}{V}\ln \mathcal{Q}(z,V,T)$$

$$n = \frac{1}{V}z\frac{\partial}{\partial z}\ln \mathcal{Q}(z,V,T). \tag{13.19}$$

For simplicity in notation, we have omitted the overhead bar indicating ensemble average, as in $n = \bar{N}/V$, since there is no danger of confusion here.

## 13.5 Critical fluctuations

We can now express the density fluctuation in terms of measurable thermodynamic coefficients. From (13.10) and (13.18), we have

$$\overline{n^2} - \bar{n}^2 == \frac{kT}{V} \frac{\partial^2 P}{\partial \mu^2}.$$

From (13.16) we obtain

$$\frac{\partial \mu}{\partial v} = -v \frac{\partial^2 a(v)}{\partial v^2} = v \frac{\partial P}{\partial v}, \tag{13.20}$$

where the dependence on $T$ is left understood. Thus

$$\frac{\partial P}{\partial \mu} = \frac{\partial P}{\partial v} \frac{\partial v}{\partial \mu} = \frac{\partial P / \partial v}{\partial \mu / \partial v} = \frac{1}{v},$$

$$\frac{\partial^2 P}{\partial \mu^2} = -\frac{1}{v^2} \frac{\partial v}{\partial \mu} = \frac{\kappa_T}{v^2}, \tag{13.21}$$

where

$$\kappa_T = -\frac{1}{v} \left( \frac{\partial v}{\partial P} \right)_T \tag{13.22}$$

is the isothermal compressibility. Thus we obtain

$$\frac{\overline{n^2} - \bar{n}^2}{\bar{n}^2} = \frac{kT\kappa_T}{V}. \tag{13.23}$$

When $V \to \infty$, this vanishes unless $\kappa_T \to \infty$, which happens at the critical point.

The strong density fluctuations at the critical point occur on a molecular scale, where the atoms come together momentarily to form large clusters, only to break loose again. The scattering cross section of light is proportional to the mean-square fluctuation of density, and becomes very large at the critical point. This gives rise to the phenomenon of *critical opalescence*. In $CO_2$, the intensity of scattered light increases a million fold at $T_c = 304$ K, $P_c = 74$ atm, and the normally transparent liquid turns milky white.

During a first-order phase transition, the pressure is independent of volume in the transition region, and one might think that $\kappa_T = \infty$; but this is not so because $\kappa_T$ refers to the compressibility of a pure phase. In the transition region, the system is a mixture, and the compressibilities of the components remain finite. It is true that the density of the mixture will fluctuate, if there is no gravity to keep the two phases apart; but such fluctuations occur on a macroscopic scale, and do not lead to opalescence. This is illustrated in Fig. 13.2.

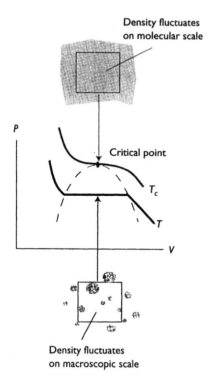

Density fluctuates
on molecular scale

Critical point

Density fluctuates
on macroscopic scale

*Figure 13.2* Critical opalescence at the critical point is due to the intense scattering of
light by density fluctuations on a molecular scale. Density fluctuations on a
macroscopic scale, such as that in a liquid–gas mixture, does not scatter light
as strongly.

## 13.6   Quantum gases in the grand canonical ensemble

For the quantum ideal gases, we can represent the energy eigenvalues in the form

$$E\{n\} = \sum_{\mathbf{k}} n_{\mathbf{k}} \epsilon_{\mathbf{k}}, \tag{13.24}$$

where $n_{\mathbf{k}}$ is the occupation number of the single-particle state with wavevector $\mathbf{k}$,
energy $\epsilon_{\mathbf{k}} = \hbar^2 k^2 / 2m$. The statistics is specified through the range $n_{\mathbf{k}} = 0, 1$ for
fermions, and $n_{\mathbf{k}} = 0, 1, \ldots, \infty$ for bosons. The canonical partition function is

$$Q_N(V, T) = \sum_{\substack{\{n\} \\ \Sigma_k n_k = N}} e^{-\beta E\{n\}}, \tag{13.25}$$

where the sum is over all possible sets of occupation numbers $\{n\}$ subject to the
condition $\sum_{\mathbf{k}} n_{\mathbf{k}} = N$. This constraint makes it impossible to calculate this sum.

Consider now the grand partition function:

$$Q(z, V, T) = \sum_{N=0}^{\infty} z^N Q_N(V, T)$$

$$= \sum_{N=0}^{\infty} \sum_{\substack{\{n\} \\ \Sigma_k n_k = N}} z^{\Sigma_q n_q} e^{-\beta \Sigma_k n_k \epsilon_k}, \tag{13.26}$$

where we have replaced the power $N$ in $z^N$ by $\Sigma_q n_q$. The double summation now collapses to a single sum over $\{n\}$ without constraint:

$$\sum_{N=0}^{\infty} \sum_{\substack{\{n\} \\ \Sigma_k n_k = N}} = \sum_{\{n\}}. \tag{13.27}$$

On the left side we first sum over the set of occupation numbers with a given $N$, and then sum over all $N$. On the right we have exactly the same sum, except that we don't bother to regroup the terms. We can further rewrite

$$z^{\Sigma_q n_q} e^{-\beta \Sigma_k n_k \epsilon_k} = \prod_q z^{n_q} \prod_k \left(e^{-\beta \epsilon_k}\right)^{n_k} = \prod_k \left(ze^{-\beta \epsilon_k}\right)^{n_k}. \tag{13.28}$$

Thus

$$Q(z, V, T) = \sum_{\{n_0, n_1, \ldots\}} z^{\Sigma_q n_q} e^{-\beta \Sigma_k n_k \epsilon_k}$$

$$= \sum_{n_0} \left(ze^{-\beta \epsilon_0}\right)^{n_0} \sum_{n_1} \left(ze^{-\beta \epsilon_1}\right)^{n_1} \cdots$$

$$= \begin{cases} \prod_k \left(1 + ze^{-\beta \epsilon_k}\right) & \text{(Fermi).} \\ \prod_k \left(1 - ze^{-\beta \epsilon_k}\right)^{-1} & \text{(Bose).} \end{cases} \tag{13.29}$$

The average occupation number is given by

$$\langle n_k \rangle = \frac{1}{Q(z, V, T)} \sum_{\{n\}} n_k z^{\Sigma_k n_k} e^{-\beta \Sigma_k n_k \epsilon_k} = -\frac{1}{\beta} \frac{\partial}{\partial \epsilon_k} \ln Q(z, V, T)$$

$$= \frac{1}{z^{-1} e^{\beta \epsilon_k} \pm 1} \quad \begin{array}{l} (+ : \text{Fermi}) \\ (- : \text{Bose}) \end{array} \tag{13.30}$$

We go to the limit $V \to \infty$ and obtain the equations of state

$$\frac{P}{kT} = \frac{1}{V} \ln Q(z, V, T) = \pm \frac{1}{2\pi^2} \int_0^{\infty} dk \, k^2 \ln \left(1 \pm ze^{-\beta \epsilon_k}\right). \tag{13.31}$$

Making a partial integration using

$$\int dk \, k^2 f(k) = \frac{k^3}{3} f(k) - \int dk \frac{k^3}{3} \frac{\partial}{\partial k} f(k), \tag{13.32}$$

we obtain

$$P = \frac{2}{3} \int \frac{d^3k}{(2\pi)^3} \frac{\epsilon_k}{z^{-1}e^{\beta \epsilon_k} \pm 1} = \frac{2}{3}U. \tag{13.33}$$

This agrees with the result obtained in the microcanonical ensemble in Section 8.9.

## 13.7  Occupation number fluctuations

To calculate the mean-square fluctuation in the average occupation number of the ideal gases, we start with the formula

$$\langle n_k \rangle = \frac{1}{Q} \sum_{\{n_k\}} n_k e^{-\beta \Sigma_k n_k (\epsilon_k - \mu)}. \tag{13.34}$$

Differentiating both sides with respect to $\epsilon_k$ for $k \neq 0$, we have

$$\frac{\partial}{\partial \epsilon_k} \langle n_k \rangle = -\frac{\beta}{Q} \sum_{\{n_k\}} n_k^2 e^{-\beta \Sigma_k n_k (\epsilon_k - \mu)} - \frac{1}{Q^2} \frac{\partial Q}{\partial \epsilon_k} \sum_{\{n_k\}} n_k e^{-\beta \Sigma_k n_k (\epsilon_k - \mu)}. \tag{13.35}$$

The last term on the right-hand side is of the form

$$-\frac{\partial \ln Q}{\partial \epsilon_k} \langle n_k \rangle = \beta \langle n_k \rangle^2. \tag{13.36}$$

Thus

$$\frac{\partial}{\partial \epsilon_k} \langle n_k \rangle = -\beta \langle n_k^2 \rangle + \beta \langle n_k \rangle^2 \tag{13.37}$$

or

$$\langle n_k^2 \rangle - \langle n_k \rangle^2 = -\frac{1}{\beta} \frac{\partial}{\partial \epsilon_k} \langle n_k \rangle$$

$$= \frac{e^{-\beta(\epsilon_k - \mu)}}{\left[1 \pm e^{-\beta(\epsilon_k - \mu)}\right]^2} \quad \begin{matrix} (-: \text{Fermi}) \\ (+: \text{Bose}) \end{matrix} \tag{13.38}$$

This can also be rewritten as

$$\langle n_k^2 \rangle - \langle n_k \rangle^2 = \langle n_k \rangle \mp \langle n_k \rangle^2 \quad \begin{matrix} (-: \text{Fermi}) \\ (+: \text{Bose}) \end{matrix} \tag{13.39}$$

The result for Fermi statistics is obvious, because $n_k^2 = n_k$. The above formulas are derived assuming $k \neq 0$. This makes little difference in the Fermi case. In the Bose case, it excludes the fluctuation of the Bose–Einstein condensate.

For a macroscopic system, the energy levels form a continuum in the thermodynamic limit. It is then more relevant to consider a group of states. It can be shown that the fluctuation in the occupation of a group of states with $k \neq 0$ is always normal (see Problem 13.11). The fluctuation of a single state is relevant only for the Bose–Einstein condensate. It can be shown that the condensate fluctuation is larger than normal (see Problem 13.12).

## 13.8 Photon fluctuations

According to (13.39), the fractional fluctuation in the occupation of a single state
for bosons is given by

$$\frac{\langle n_k^2 \rangle - \langle n_k \rangle^2}{\langle n_k \rangle^2} = \frac{1}{\langle n_k \rangle^2} + 1, \tag{13.40}$$

where $k \neq 0$. This is always greater than 100%, and was experimentally observed
by Hanbury-Brown and Twiss (Hanbury-Brown 1957) in an almost monochromatic
beam of photons from a thermal source.

Their instrument is schematically illustrated in Fig. 13.3. A half-silvered mirror
splits the light beam in two, and the intensity of one of the beams is measured at
time $t$, and the other is measured at time $t + \tau$. The product of the intensities is
then averaged over $t$. This is equivalent to measuring the intensity of a single beam
at two different points separated by a distance $c\tau$ simultaneously, and averaging
the product of the intensities over time.

We shall not go into quantitative details, but just report that photons were found
to arrive in bunches. Since the intensity at a particular location gives the photon
occupation number in the thermal source at an earlier time, the bunching of photons
indicates that the occupation number strongly fluctuates in time.

Purcell (1956) has given an intuitive explanation why monochromatic photons
fluctuate in number. Think of a photon beam in a cavity bouncing between walls,
or one that came out of the cavity. In the real world, no beam can be exactly
monochromatic, for that would correspond to a plane wave that fills all space. The
photon beam therefore consists of a stream of random wave trains, each containing
one photon, with length of order $c/\Delta \nu$, where $\Delta \nu$ is the uncertainty in frequency.
For small $\Delta \nu$ the wave train can be very long. Occasionally two trains will overlap,

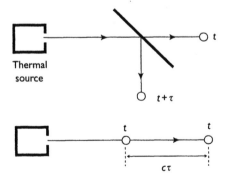

*Figure 13.3* In the experiment of Hanbury-Brown and Twiss, a light beam from a thermalized source is split, and the intensities are measured at the two ends with time
difference $\tau$. This is equivalent to taking instantaneous measurements along an
unsplit beam at a distance $c\tau$ apart, where $c$ is the velocity of light.

and when that happens one gets either 0 or 4 photons. This is why the photons appear bunched together.

The phenomenon observed can be quantitatively explained in terms of classical electromagnetic waves, and it has been questioned whether it has anything to do with Bose statistics. The answer is yes, because the Bose statistics allows many photons to be in the same state, which leads to a classical electromagnetic wave when the number is large enough.

It should be mentioned that a laser beam does not show fluctuations in intensity. That's because a laser is not in a thermalized state; it is a coherent superposition of states with different numbers of photons.

## 13.9  Pair creation

The grand canonical ensemble includes systems with different particle numbers, with a mean value $N$ determined by the chemical potential. This makes sense only if $N$ is a conserved quantity, for otherwise the chemical potential would be zero, as in the case of photons. In the everyday world, the truly conserved quantity is $N - \bar{N}$, where $N$ and $\bar{N}$ are, respectively, the number of atoms and antiatoms. It appears that $N$ is conserved only because there are no antiatoms around at room temperature.

We consider an example in which both particle and antiparticle are important. This is the case of electrons and positrons in the interior of a star, which can be pair-created and pair-annihilated through the reaction

$$e^+ + e^- \rightleftarrows \text{radiation}, \tag{13.41}$$

where the radiation consists of photons with a Planck distribution at a very high temperature. The reaction establishes an average value for the conserved quantum number $N_+ - N_-$, where $N_\pm$ is the number of positrons and electrons, respectively. To find the equilibrium condition, it is not necessary to know the transition rate, which determines how long it will take to establish equilibrium.

The grand partition function is given by

$$\mathcal{Q} = \sum_{N_+=0}^{\infty} \sum_{N_-=0}^{\infty} z^{N_+-N_-} \mathcal{Q}_{N_+} \mathcal{Q}_{N_-}, \tag{13.42}$$

where $\mathcal{Q}_N$ is the partition function for a free electron or positron gas. Writing $\mathcal{Q}_N = e^{-\beta A_N}$, and $z = e^{\beta \nu}$, we have

$$\ln \mathcal{Q} = \ln \sum_{N_+=0}^{\infty} \sum_{N_-=0}^{\infty} \exp\left\{ -\beta \left[ A_{N_+} + A_{N_+-} - \nu \left( N_+ - N_- \right) \right] \right\}. \tag{13.43}$$

Assuming that the fluctuations of $N_\pm$ are small, we keep only the largest term in the sum, which is determined by the conditions

$$\frac{\partial}{\partial N_+} A_{N_+} = \nu,$$

$$\frac{\partial}{\partial N_-} A_{N_-} = -\nu. \tag{13.44}$$

Thus, the chemical potentials of the two gases must be equal and opposite:

$$\mu_+ + \mu_- = 0, \tag{13.45}$$

where $\mu = \partial A_N / \partial N$. This will determine the equilibrium ratio of electrons and positrons.

Assuming $kT \ll mc^2$, we use the non-relativistic limit (See Problem 12.7)

$$\mu = kT \ln\left(n\lambda^3\right) + mc^2, \tag{13.46}$$

where $\lambda = \sqrt{2\pi\hbar^2/mkT}$ is the thermal wavelength. It is important to keep the rest energy $mc^2$, because we are considering reactions that convert mass into energy and vice versa. The condition for equilibrium is then

$$kT\left[\ln\left(n_+\lambda^3\right) + \ln\left(n_-\lambda^3\right)\right] + 2mc^2 = 0, \tag{13.47}$$

where $n_\pm = N_\pm / V$. We can rewrite the formula as

$$n_+ n_- = \lambda^{-6} e^{-2mc^2/kT}. \tag{13.48}$$

Assuming the initial value

$$n_- - n_+ = n_0, \tag{13.49}$$

where $n_0 > 0$, we obtain

$$n_+^2 + n_0 n_+ - \lambda^{-6} e^{-2mc^2/kT} = 0. \tag{13.50}$$

For $n_+/n_0 \ll 1$, the solutions are

$$\frac{n_+}{n_0} \approx \frac{1}{\lambda^6 n_0^2} e^{-2mc^2/kT},$$

$$\frac{n_-}{n_0} \approx 1 + \frac{1}{\lambda^6 n_0^2} e^{-2mc^2/kT}. \tag{13.51}$$

## Problems

**13.1**   A lattice gas consists of $N_0$ sites, each of which may be occupied by at most one atom. The energy of a site is $\epsilon$ if occupied, and 0 if empty. The atoms are indistinguishable.

(a)  Calculate the grand partition function $\mathcal{Q}(z, T)$ at fugacity $z$ and temperature $T$.
(b)  What fraction of the sites are occupied?
(c)  Find the heat capacity as a function of $T$ at fixed $z$.

**13.2**   Carbon-monoxide poisoning happens when CO replaces $O_2$ on Hb (hemoglobin) molecules in the blood stream. Consider a model of Hb consisting of $N$ sites, each of which may be empty (energy 0), occupied by $O_2$ (energy $\epsilon_1$), or occupied by CO (energy $\epsilon_2$). At body temperature 37°C, the fugacities of $O_2$ and CO are, respectively, $z_1 = 10^{-5}$ and $z_2 = 10^{-7}$.

(a)  Consider first the system in the absence of CO. Find $\epsilon_1$ (in eV) such that 90% of the Hb sites are occupied by $O_2$.
(b)  Now admit CO. Find $\epsilon_2$ (in eV) such that 10% of the sites are occupied by $O_2$.

**13.3**   Gas molecules can adsorb on the surface of a solid at $N$ possible adsorption sites. Each site has binding energy $\epsilon$, and can accommodate at most one molecule. The adsorbed molecules are in equilibrium with a gas surrounding the solid (see sketch). We can treat the system of adsorbed molecules in a grand canonical ensemble with temperature $T$ and chemical potential $\mu$.

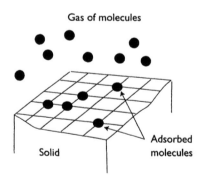

Gas of molecules

Solid

Adsorbed
molecules

(a)  If there are $M$ adsorbed molecules, what is the energy $E(M)$ of the system? What is the degeneracy $\Gamma(M)$ of the energy?
(b)  Write down the grand partition function of the system as a sum over $M$. Determine the thermal average $\bar{M}$ as the value that maximized the summand.
(c)  Suppose the gas surrounding the solid has pressure $P$. Calculate $\bar{M}$ using the ideal-gas expression for $\mu$.
(d)  Find $\overline{M^2} - \bar{M}^2$.

**13.4** (a) Calculate the isothermal compressibility of a van der Waals gas near the critical volume, with $T \rightarrow T_c$ from above. Use the reduced form of the equation of state, with the critical point located at $P = V = T = 1$.

(b) Describe how the density fluctuation diverges when $T \rightarrow T_c$ from above.

**13.5** In Section 13.9 we worked out the equilibrium distribution of electrons and positrons in the low-temperature limit $kT \ll mc^2$. Solve the problem in the ultra-relativistic limit. In the high-temperature limit $kT \ll mc^2$. The chemical potential in this limit is given in Problem 12.7:

$$\mu \approx kT \ln \left( nL^3 \right),$$

$$L = \frac{\pi^{2/3} \hbar c}{kT}.$$

**13.6** A chemical reaction, such as $2H_2 + O_2 \rightleftharpoons 2H_2O$, can be denoted in the form

$$v_1 X_1 + v_2 X_2 + \cdots \rightleftharpoons v_1' Y_1 + v_2' Y_2 + \cdots$$

Taking $v_i' = -v_i$, we can rewrite this as

$$\sum_i v_i X_i = 0.$$

The numbers $v_i$ are called *stoichiometric coefficients*. Consider a mixture of $N_i$ molecules of the type $X_i$.

(a) Show that in a chemical reaction the changes $\delta N_i$ in the numbers satisfy the relation

$$\frac{\delta N_1}{v_1} = \frac{\delta N_2}{v_2} = \cdots$$

Thus, $\delta N_i / v_i$ is independent of $i$, and we can put $\delta N_i = v_i \delta N$, where $\delta N$ is some constant.

(b) Assume that the total Helmholtz free energy of the mixture is the sum of those of the components. Minimize the free energy by varying $N_i$ at constant volume and temperature, and show that in thermodynamic equilibrium

$$\sum_i \mu_i v_i = 0,$$

where $\mu_i$ is the chemical potential of the $i$th component.

**13.7** Apply the results of the last problem to $2H_2 + O_2 \rightleftharpoons 2H_2O$, treating the components as classical ideal gases. Number the components 1, 2, 3. The masses are $m_1 = 2m$, $m_2 = 32m$, $m_3 = 18m$, where $m$ is the nucleon mass. The stoichiometric coefficients are $v_1 = 2$, $v_2 = 1$, $v_3 = -2$.

(a) Show that there are two conservation laws: $n_1 - 2n_2 = A$, and $n_1 + n_3 = B$, where $A, B$ are constants.

(b) Assume that initially there was no $H_2O$, that $H_2$ and $O_2$ were present in the ration 2:1, and that the density of $H_2$ was $n_0$. Find the equation determining $n_1/n_0$ as a function of temperature, and solve it in the high-temperature and low-temperature limits. Give the results for $n_2/n_0$ and $n_3/n_0$.

**13.8**   The equation of state of a $P$–$V$–$T$ system can be written in the parametric form

$$\frac{P}{kT} = \frac{1}{\lambda^3} \sum_{\ell=1}^{\infty} b_\ell z^\ell,$$

$$\frac{N}{V} = \frac{1}{\lambda^3} \sum_{\ell=1}^{\infty} \ell b_\ell z^\ell,$$

where $\lambda = \sqrt{2\pi\hbar^2/mkT}$, and $b_\ell$ are coefficients known as "cluster integrals", with $b_1 \equiv 1$.

(a) Find $b_\ell$ for the ideal Bose gas and the ideal Fermi gas.
(b) For low densities, we can expand the pressure in a power series in the density $n = N/V$ called the virial expansion:

$$\frac{PV}{NkT} = 1 + a_2 \left(\lambda^3 n\right) + a_3 \left(\lambda^3 n\right)^2 + \cdots,$$

where $a_\ell$ is called the $\ell$th virial coefficient. They can be expressed in terms of the cluster integrals by eliminating $z$ from the parametric equation of state. Show that

$$a_2 = -b_2,$$

$$a_3 = 4b_2^2 - 2b_3.$$

**13.9**   Consider an ideal Fermi gas of $N$ spinless particles, with single-particle states labeled by a set of quantum numbers $\{\lambda\}$. Denote the energy levels by $\epsilon_\lambda$.

(a) Suppose the energy spectrum consists of a bound state of energy $-B$ that is $g$-fold degenerate, plus the usual free-particle energy spectrum $\epsilon_k = \hbar^2 k^2/2m$. Assume $g = aV$, where $V$ is the volume, and $a$ is a constant. Give the equation of state in parametric form.
(b) Find $z$ at high temperatures, as a function of temperature $T$ and density $n$, to lowest order.
(c) In the same approximation, find the pressure $P$, and the densities $n_b$ and $n_f$ of bound and free particles, respectively.

**13.10**   For an ideal quantum gas show that

$$\langle n_k n_p \rangle - \langle n_k \rangle \langle n_p \rangle = -\frac{1}{\beta}\frac{\partial}{\partial \epsilon_k}\langle n_p \rangle \quad (k \neq p).$$

Since $\langle n_p \rangle$ does not depend on $\epsilon_k$, this gives

$$\langle n_k n_p \rangle - \langle n_k \rangle \langle n_p \rangle = 0 \quad (k \neq p).$$

(*Hint:* Differentiate the grand partition function with respect to $\epsilon_p$, and then $\epsilon_k$.)

**13.11**   We have obtained the fluctuation for the occupation of a single state. For a macroscopic system, the energy levels form a continuum, and it is more relevant to consider a group of states. Let $\sigma$ be the occupation number of a group $G$:

$$\sigma = \sum_{k \in G} n_k.$$

Show, with help from the result of the last problem,

$$\langle \sigma^2 \rangle - \langle \sigma \rangle^2 = \langle \sigma \rangle \mp \sum_{k \in G} \langle n_k \rangle^2 \quad (-: \text{Fermi})(+: \text{Bose})$$

In the limit $V \to \infty$, the sum $\sum_k$ approaches an integral $V(2\pi)^{-3}\int d^3k$. If no level is macroscopically occupied, then the right-hand side is of order $V$, while the left-hand side is of order $V^2$. In that case, the fluctuation is normal. This is always true for the Fermi case, and is also true for the Bose case in the gas phase.

**13.12**   *Condensate fluctuation*: The formulas for occupation-number fluctuations given so far have excluded the zero-momentum state. The missing information is relevant for an ideal Bose gas in the region of Bose–Einstein condensation. Show that the condensate fluctuation is above normal, as follows.

(a) Show that the number of particles in the condensate has a mean-square fluctuation

$$\langle n_0^2 \rangle - \langle n_0 \rangle^2 = \sum_{k \neq 0}\left[\langle n_k^2 \rangle - \langle n_k \rangle^2\right].$$

(*Hint:* Use the definition $n_0 = N - \sum_{k \neq 0} n_k$.)

(b) Show that the fluctuation is proportional to $V^{4/3}$ when $T < T_c$. (*Hint:* In the thermodynamic limit $V \to \infty$, the sum on the right-hand side diverges at $z = 1$. Keep $V$ large but finite, and estimate the asymptotic $V$ dependence. To do this, you must consider how the energy levels merge into a continuum for large $V$. Use periodic boundary conditions.)

# Chapter 14

# The order parameter

## 14.1 Broken symmetry

Nature does not always favor the most symmetrical arrangement, contrary to metaphysical expectations. Aristotle held that planetary orbits must be circles because they were "perfect" – apparently they are not. Another example is given in Problem 14.1.

As a rule, matter at low temperatures goes into a phase that does not possess the intrinsic symmetry of the system. Atoms solidify into a crystal lattice, which violates the translational invariance of the Hamiltonian. The atomic spins in a ferromagnetic material line up along a definite direction, violating rotational invariance. When the Hamiltonian is invariant under a certain symmetry operation, and the ground state is not invariant, that symmetry is said to be "spontaneously broken".

Suppose $P$ is a symmetry operation under which the Hamiltonian $H$ is invariant, and $\Psi_0$ is a ground state wave function. We have

$$H\Psi_0 = E_0\Psi_0,$$
$$P^{-1}HP = H. \tag{14.1}$$

This means that

$$P^{-1}HP\Psi_0 = E_0\Psi_0,$$
$$H(P\Psi_0) = E_0(P\Psi_0). \tag{14.2}$$

If $P\Psi_0$ is a different state from $\Psi_0$, it is an equally valid ground state. Therefore, if the symmetry is broken, the ground state must be degenerate.

In the breaking of a continuous symmetry, the ground state must be infinitely degenerate. For a ferromagnet, the total spin of the system must point along a definite direction, but it can be any direction in space. A consequence of this degeneracy is the existence of *spin wave* excitations, in which the spin direction varies slightly from point to point in space, as illustrated in Fig. 14.1. In the long wavelength limit, the excitation energy is vanishingly small, because locally the system is in a possible ground state. This is call a "Goldstone mode", the signature of a broken continuous symmetry (See Huang 1998, Chapter 15).

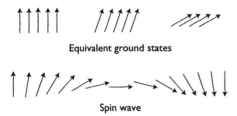

Equivalent ground states

Spin wave

*Figure 14.1* Broken rotational symmetry: A ferromagnet has an infinite number of degenerate ground states, although the system can be in only one of them. The symmetry is expressed through the existence of spin waves, in which neighboring points in this system are in slightly different ground states.

The physical origin of spontaneous magnetization is the attraction between parallel spins. They prefer to be aligned and rotate as a block. However, this tendency towards organization is opposed by random thermal fluctuations, and the balance between these opposite forces determines the average block size at a given temperature. At high temperatures, the block size is small, their random spin orientations cancel each other, and there is no net magnetization. As the temperature is decreased, larger blocks become more stable, and it becomes increasingly rare that the block-spin direction will change as a result of thermal motion. This is because changing the block-spin direction requires the spontaneous cooperative rotation of the individual spins, and this is very unlikely for large blocks. Broken symmetry comes about because the system gets trapped in some configuration for a very long time. That is, it originates from a failure of ergodicity.

The competition between spin alignment and thermal motion can be seen in the free energy $A = U - TS$, where $U$ prefers minimization through spin alignment, whereas $S$ seeks maximization by randomizing the spin orientations. When $T$ is decreased, the forces of alignment gain favor, and win at $T = T_c$, where $U - T_cS = 0$. The emergence of spontaneous magnetization gives rise to a sharp phase transition in the thermodynamic limit. The rotational symmetry is broken in the low-temperature phase, and "restored" in the high-temperature phase. The main subject of this chapter is a phenomenological theory of such a phase transition, which is generally of second order.

From the spontaneous magnetization we abstract the notion of an "order parameter", whose emergence below a critical temperature marks a symmetry-breaking phase transition. Some physical examples of the order parameter are given in Table 14.1.

## 14.2 Ising spin model

We illustrate spontaneous symmetry breaking in a simple model of ferromagnetism, the Ising spin model. It consists of a lattice of two-valued spins $s_i = \pm 1$

Table 14.1 Order parameters

| System | Order parameter | Symmetry broken |
|---|---|---|
| Ferromagnet | Magnetization | Rotational invariance |
| Antiferromagnet | Staggered magnetization | Rotational invariance |
| Coexisting liquid–gas | Density difference | Spatial homogeneity |
| Superfluid | Condensate wave function | Global gauge invariance |

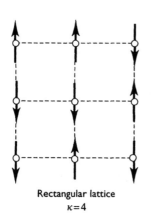

Rectangular lattice
$\kappa=4$

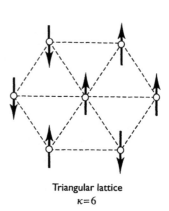

Triangular lattice
$\kappa=6$

Figure 14.2 Examples of Ising spin model, in 2D. $\kappa$ denotes the number of nearest neighbors.

in a $D$-dimensional space. The number of nearest neighbors to any given site is denoted by $\kappa$. Some 2D examples are shown in Fig. 14.2. We assume that the total number of sites $N$ is so large that physical results are independent of the boundary conditions.

A spin configuration is denoted by $\{s\} = \{s_1, s_2, \ldots, s_N\}$, with energy given by

$$E\{s\} = -\epsilon \sum_{\langle ij \rangle} s_i s_j - h \sum_i s_i, \tag{14.3}$$

where $\langle ij \rangle$ denotes a nearest-neighbor pair of sites, and $h$ is a uniform external magnetic field. The magnetic moment of a spin is unity in our units. The constant $\epsilon$ is positive for a ferromagnet, and negative for an antiferromagnet. We shall assume $\epsilon > 0$. The canonical ensemble average of the total spin is the magnetization

$$M(h) = \frac{\sum_{\{s\}} e^{-\beta E\{s\}} \sum_i s_i}{\sum_{\{s\}} e^{-\beta E\{s\}}}, \tag{14.4}$$

where $\beta = 1/kT$. We call this the order parameter.

When $h = 0$ there is up-down invariance: $E\{s\} = E\{-s\}$. Thus, it seems that spontaneous magnetization is impossible, for, by changing the variable of summation from $\{s\}$ to $\{-s\}$, we find $M(0) = -M(0)$, and therefore $M(0) = 0$. The resolution of the paradox lies in the argument we gave earlier: in the limit $N \to \infty$, at sufficiently low temperatures, the system cannot sample all available configurations in finite time. Mathematically, this is expressed through the fact that, at sufficiently low temperatures, the limits $h \to 0$, and $N \to \infty$ do not commute:

$$\lim_{N\to\infty} \lim_{h\to0} M(h) = 0,$$

$$\lim_{h\to0} \lim_{N\to\infty} M(h) \neq 0. \tag{14.5}$$

The second limit corresponds to spontaneous symmetry breaking.

We shall find $M(h)$ using an approximate method. The energy of a particular spin $s_i$ is given by

$$w(s_i) = -s_i h - \epsilon s_i \sum_{\mathrm{nn}\ i} s_j, \tag{14.6}$$

where nn means "nearest-neighbor to". The *mean-field approximation* consists of replacing each nearest-neighbor spin by the average spin per particle, denoted by

$$m(h) = \frac{M(h)}{N}. \tag{14.7}$$

Thus, the sum over nearest-neighbor spins yields a factor $\kappa m$:

$$w(s_i) = -[h + \epsilon\kappa m]\, s_i. \tag{14.8}$$

The spin $s_i$ sees an effective magnetic field

$$h_{\mathrm{eff}} = h + \kappa\epsilon m. \tag{14.9}$$

In thermal equilibrium, let there be $N_+$ up spins, and $N_-$ down spins in the system. Their probabilities must be given by

$$\frac{N_\pm}{N} = \frac{e^{\pm\beta h_{\mathrm{eff}}}}{e^{\beta h_{\mathrm{eff}}} + e^{-\beta h_{\mathrm{eff}}}}. \tag{14.10}$$

Thus, the average spin is

$$m = \frac{N_+ - N_-}{N} = \frac{e^{\beta h_{\mathrm{eff}}} - e^{-\beta h_{\mathrm{eff}}}}{e^{\beta h_{\mathrm{eff}}} + e^{-\beta h_{\mathrm{eff}}}}$$

$$= \tanh(\beta h_{\mathrm{eff}}). \tag{14.11}$$

Using (14.9), we obtain the self-consistency condition

$$m = \tanh(\beta(h + \kappa\epsilon m)), \tag{14.12}$$

which can be solved graphically, by plotting the left-hand sides and right-hand sides on a graph, and finding their intersections.

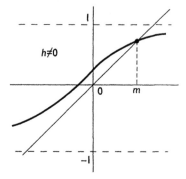

Figure 14.3 Graphical solution for the magnetic moment per particle, $m$, in the mean-field theory for nonzero external field. There is always one root representing induced magnetization.

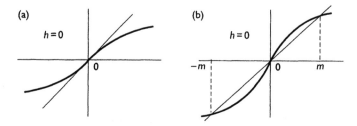

Figure 14.4 For zero external field, spontaneous magnetization occurs when the slope of the curve at the origin exceeds unity.

In Fig. 14.3, we show the graphical solution for the case $h \neq 0$. The plot of the right-hand side does not vanish at the origin, and there is always one and only one root, giving an induced magnetic moment.

In Fig. 14.4, we illustrate the case $h = 0$. The plot of the right-hand side is symmetric under a reflection about the origin. If the slope at the origin is less than unity, as shown in panel (a), there is no root other than $m = 0$, and hence no spontaneous magnetization. When the slope is greater than unity, as shown in panel (b), there are two equal and opposite roots $\pm m$. There is still an intersection at $m = 0$, but that gives a higher energy. We must choose one of the two roots, thereby breaking the up–down symmetry. The condition for this to happen is $\beta \gamma \epsilon > 1$, and the critical temperature $T_c$ is given by

$$kT_c = \kappa \epsilon. \tag{14.13}$$

The magnetization as a function of temperature looks like that in Fig. 1.5 in Chapter 1. There is no spin wave in this model, because the broken symmetry is discrete instead of continuous.

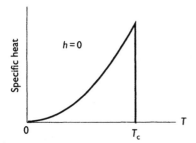

Figure 14.5 Specific heat of Ising spin model in the mean-field approximation.

In the mean-field approximation, the internal energy of the particle is

$$U = -Mh_{\text{eff}} = -Nm(h + kT_c m). \tag{14.14}$$

The specific heat at constant external field at $h = 0$ is given by

$$C = \frac{1}{N}\left(\frac{\partial U}{\partial T}\right)_{h=0} = -2kT_c m \frac{\partial m}{\partial T}. \tag{14.15}$$

Its qualitative behavior is depicted in Fig. 14.5.

The mean-field approximation is not correct in details. It predicts a phase transition independent of $D$, but actually there is no phase transition for $D = 1$. Nevertheless, it illustrates in a simple way how spontaneous magnetization can suddenly arise, in a theory dealing only with continuous functions. This idea inspires a more elaborate theory of the order parameter, which we discuss next.

## 14.3  Ginsburg–Landau theory

The Ginsburg–Landau theory is based on the view that second-order phase transitions have universal properties, in that they depend only on the nature of the order parameter, independent of the finer details of the system.

We ignore all the microscopic details of the system, and describe it only through a field $\phi(x)$ in $D$-dimensional space, which is modeled after the magnetization density. The statistical ensemble describing the system is the collection of all possible functional forms $\phi$, weighted according to a Boltzmann factor $e^{-\beta E[\phi]}$, where $E[\phi]$ is a functional of $\phi$, called the *Landau free energy*, postulated to have the form

$$E[\phi] = \int d^D x \left[\frac{1}{2}|\nabla\phi(x)|^2 + \Omega(\phi(x)) - h(x)\phi(x)\right]. \tag{14.16}$$

Note that there is a kinetic term $|\nabla\phi|^2$, a potential $\Omega(\phi)$, and an external field $h(x)$. The kinetic term imposes an energy cost for a gradient in $\phi$, and drives the system towards uniformity. The potential is chosen as a power series in $\phi$ with even powers:

$$\Omega(\phi(x)) = r_0\phi^2(x) + u_0\phi^4(x) + \cdots \qquad (14.17)$$

This gives the system an "up–down" symmetry, with respect to $\phi \to -\phi$. The parameters $r_0$ and $u_0$ are functions of the temperature. We require $u_0 > 0$, so that $E[\phi]$ has a lower bound; but $r_0$ can have either sign. In fact, a phase transition occurs when $r_0$ changes sign. In a more general model, we can consider a multi-component $\phi$, for example a complex number, or a vector quantity.

Note that a functional such as $E[\phi]$ is a number whose value depends on the functional form of $\phi$. In contrast, a function such as $\Omega(\phi(x))$ is a number whose value depends on the value of $\phi$ at $x$.

Our object is to describe the phase transition corresponding to the breaking of the up–down symmetry. The symmetry-breaking phase transition happens when the ensemble average of $\phi$ becomes nonzero. Thus, the important configurations for $\phi$ are small near the transition point, and we keep only the first two non-vanishing terms in the potential.

The partition function is obtained by integrating over all possible functional forms of $\phi$:

$$Q[h] = \int (D\phi)\, e^{-\beta E[\phi]}, \qquad (14.18)$$

where $\beta = 1/kT$, and $\int (D\phi)$ denotes functional integration. We imagine that this formula is the result of summing over microscopic configurations with the same $\phi$, and that the entropy factor arising from this process has been incorporated into the definition of $E[\phi]$ – hence the name "free energy". The thermodynamic free energy is given by

$$A[h] = -kT \ln Q[h] \qquad (14.19)$$

and the ensemble average of a quantity $O$ is given by

$$\langle O \rangle = \frac{\int (D\phi)\, O\, e^{-\beta E[\phi]}}{\int (D\phi)\, e^{-\beta E[\phi]}}. \qquad (14.20)$$

The functional integration can be performed as follows. We replace the continuous space of $x$ by a discrete lattice of points $\{x_1, x_2, \ldots\}$. Let us denote $\phi_i = \phi(x_i)$. A functional integral over $\phi$ can be approximated by the multiple integral over the independent values $\{\phi_1, \phi_2, \ldots\}$:

$$\int (D\phi) = \int_{-\infty}^{\infty} d\phi_1 \int_{-\infty}^{\infty} d\phi_2 \cdots \qquad (14.21)$$

This is illustrated in Fig. 14.6. We approach the continuum limit by making the lattice spacing smaller and smaller with respect to some fixed scale. In the limit,

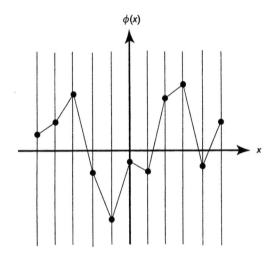

*Figure 14.6* The functional integral over $\phi$ can be approximated by integrating over the values of $\phi(x_i)$, independently at a discrete set of $x_i$. The continuum limit is taken after a physical quantity is calculated with this method.

the functional integration includes continuous as well as discontinuous functions. In fact, the vast majority of the functions are discontinuous, but because of their large kinetic contribution, they give small contributions to the partition function. The continuum limit of the functional integral itself may be ambiguous, but it usually exists for an ensemble average.

The order parameter is taken to be

$$M = \left\langle \int d^D x \phi(x) \right\rangle,$$  (14.22)

which is analogous to magnetization. For a uniform external field $h$ we can obtain $M$ through

$$M = -\frac{\partial A}{\partial h}.$$  (14.23)

The susceptibility is

$$\chi = \frac{1}{V} \frac{\partial M}{\partial h},$$  (14.24)

where $V$ is the total volume of the system. The heat capacity at constant external field is given by

$$C = -T \frac{\partial^2 A}{\partial T^2}.$$  (14.25)

## 14.4  Mean-field theory

We shall not discuss the full theory involving functional integrals, but go straight to the mean-field approximation. In a uniform external field $h$, the field $\phi$ fluctuates about a constant. In the mean-field approximation we ignore the fluctuations and put

$$\phi(x) = m, \tag{14.26}$$

which is the order parameter density. The Landau free energy now becomes a function $E(m)$, which is also the thermodynamic free energy, and will be denoted by $A(m)$:

$$A(m) = V \left( r_0 m^2 + u_0 m^4 - hm \right). \tag{14.27}$$

Minimizing this leads to the equation

$$2r_0 m + 4u_0 m^3 - h = 0. \tag{14.28}$$

For $h = 0$ we have

$$m \left( m^2 + \frac{r_0}{2u_0} \right) = 0 \tag{14.29}$$

with the possible roots $0, \pm\sqrt{-r_0/2u_0}$. Since $m$ must be real, the second pair of roots is acceptable only if $r_0 < 0$. The potential $\Omega(m)$ has the behavior shown in Fig. 14.7, which exhibits minima at

$$m = \begin{cases} 0 & (r_0 > 0), \\ \pm\sqrt{-r_0/2u_0} & (r_0 < 0). \end{cases} \tag{14.30}$$

For $r_0 < 0$, the root $m = 0$ is a maximum, and the system must choose between the two degenerate minima, thus breaking the symmetry.

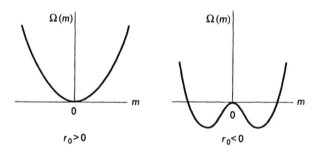

Figure 14.7  Model of a second-order symmetry-breaking phase transition. The potential develops two equivalent minima when the parameter $r_0$ changes sign.

To model a second-order symmetry-breaking phase transition, we choose

$$r_0 = bt, \tag{14.31}$$

where $b$ is a positive real constant, and

$$t = \frac{T}{T_c} - 1. \tag{14.32}$$

The parameter $u_0$ is chosen to be independent of $t$. Then we have

$$m = \begin{cases} 0 & (t > 0), \\ m_0\sqrt{t} & (t < 0), \end{cases} \tag{14.33}$$

where $m_0 = \sqrt{b/2u_0}$. This behavior agrees with the mean-field theory of the Ising spin model, in the neighborhood of $t = 0$ (see Problem 14.2). We have assumed that the system chooses the positive minimum instead of the negative one. We can tilt the system in favor of this choice by keeping an infinitesimally small positive external field while approaching $t \to 0^-$.

Note that the model works with continuous functions, and yet we get a second-order phase transition with discontinuous properties such as the specific heat. This comes about because the minimum point of a smooth curve can jump to a new value discontinuously, when the curve is continuously deformed. This is the insight underlying "catastrophe theory" (Thom 1972, Zeeman 1977).

## 14.5  Critical exponents

At the critical point, a thermodynamic function generally contains a term regular in $t$, plus a singular part that behaves like a power of $t$. The power is called the *critical exponent*. The following exponents $\alpha, \beta, \gamma$ are defined at $h = 0$ as $t \to 0$:

$$M \sim |t|^\beta \quad \text{(Order parameter)},$$

$$\chi \sim |t|^{-\gamma} \quad \text{(Susceptibility)}, \tag{14.34}$$

$$C \sim |t|^{-\alpha} \quad \text{(Heat capacity)},$$

where $\sim$ means "singular part is proportional to". (Note that $\beta$ here is not $1/kT$.) These exponents should be the same whether we approach the critical point from above or below; but the proportionality constant can be different, and may be zero on one side. Another exponent $\delta$ is defined at $t = 0$ as $h \to 0^+$:

$$M \sim h^{1/\delta} \quad \text{(Equation of state)}. \tag{14.35}$$

Critical exponents are interesting because they are universal, being shared by a class of systems. An indication of this is the fact that the Ginsburg–Landau theory can describe a wide class of physical systems, but has set of critical exponents dependent only on the dimensionality of space, the number of components of the order parameter (which is one in the model we are considering) and the number of terms in the potential.

Let us calculate the critical exponents in the mean-field theory. From (14.33) we immediately obtain

$$\beta = \frac{1}{2}. \tag{14.36}$$

At $t = 0$ and $h > 0$, we substitute the value $r = 0$ into (14.28), and obtain $4u_0m^3 - h = 0$. Therefore

$$\delta = 3. \tag{14.37}$$

To calculate the susceptibility $\chi$, we differentiate both sides of (14.28) with respect to $h$, obtaining

$$\chi = \frac{1}{2bt + 12u_0m^2}. \tag{14.38}$$

Substituting (14.33) into the right-hand side, we have

$$\chi = \begin{cases} (2bt)^{-1} & (t > 0), \\ (6bt)^{-1} & (t < 0). \end{cases} \tag{14.39}$$

Hence

$$\gamma = 1. \tag{14.40}$$

For $h = 0$ and $t \to 0$, we have from (14.27) and (14.33)

$$\frac{A}{V} = \begin{cases} 0 & (t > 0), \\ \left(-b^2/4u_0\right)t^2 & (t < 0), \end{cases} \tag{14.41}$$

from which we obtain

$$\frac{C}{V} = \begin{cases} 0 & (t > 0), \\ b^2T_c/(2u_0) & (t < 0). \end{cases} \tag{14.42}$$

This gives

$$\alpha = 0. \tag{14.43}$$

For comparison, we list the critical exponents for various systems in Table 14.2. The exponent $v$ is to be defined in a later next section. As we suspect, the mean-field approximation is deficient, as evidenced by the fact that its critical exponents are independent of $D$.

Table 14.2 Experimental values of critical exponents from various systems

| System | $\alpha$ | $\beta$ | $\gamma$ | $\delta$ | $\nu$ |
|---|---|---|---|---|---|
| Mean-field theory | 0 | $\frac{1}{2}$ | 1 | 3 | $\frac{1}{2}$ |
| 2D Ising model (exact) | 0 | $\frac{1}{8}$ | $\frac{7}{4}$ | 15 | 1 |
| 3D Ising model (approx) | 0.12 | 0.31 | 1.25 | 5 | 0.64 |
| Experiments ($D = 3$) | 0–0.14 | 0.32–0.39 | 1.3–1.4 | 4–5 | 0.6–0.7 |

## 14.6 Fluctuation–dissipation theorem

We define the correlation function by

$$G(x, y) = \langle \phi(x)\phi(y) \rangle - \langle \phi(x) \rangle \langle \phi(y) \rangle. \tag{14.44}$$

If there is no correlation between the values of the field at $x$ and $y$, then the joint average $\langle \phi(x)\phi(y) \rangle$ would be the same as the product of the individual averages, and we would have $G(x, y) = 0$. In general, the correlation is not zero, but approaches zero with increasing distance $|x - y|$.

In a uniform external field, the system is translationally invariant, and $G$ depends only on $x - y$. For the same reason, $\langle \phi(x) \rangle = \langle \phi(0) \rangle$. Thus we consider

$$G(x) = \langle \phi(x)\phi(0) \rangle - \langle \phi(0) \rangle^2. \tag{14.45}$$

According to (14.22), the order parameter is given by

$$M = \frac{V \int (D\phi)\, e^{-\beta E[\phi]} \phi(0)}{\int (D\phi) e^{-\beta E[\phi]}} \tag{14.46}$$

where we have replaced $\int d^D x\, \phi(x)$ by $V\phi(0)$. Differentiating both sides with respect to $h$, we obtain

$$\frac{1}{V}\frac{\partial M}{\partial h} = \frac{\beta \int (D\phi)\, e^{-\beta E[\phi]} \phi(0) \int d^D x\, \phi(x)}{\int (D\phi)\, e^{-\beta E[\phi]}}, \tag{14.47}$$

which can be rewritten as

$$\chi = \frac{1}{kT} \int d^D x\, G(x). \tag{14.48}$$

This is the *fluctuation–dissipation theorem*. The dissipation refers to the susceptibility on the left-hand side, and the fluctuation refers to the correlation function on the right-hand side. This theorem occurs in various forms in different contexts. We shall discuss its microscopic basis in Section 16.8.

## 14.7  Correlation length

To get an idea of how correlations decay with distance, we calculate the mean field $m(x)$ in the presence of an external field concentrated at one point:

$$h(x) = h_0 \delta^D(x). \tag{14.49}$$

We can find an equation for $m(x)$ by minimizing the Landau free energy

$$E[\phi] = \int d^D x \left[ \frac{1}{2} |\nabla \phi(x)|^2 + r_0 \phi^2(x) + u_0 \phi^4(x) - h_0 \phi(x) \delta^D(x) \right]. \tag{14.50}$$

The term $|\nabla \phi(x)|^2$ may be replaced by $-\frac{1}{2}\phi(x)\nabla^2\phi(x)$, by performing a partial integration. Put

$$\phi(x) = m(x) + \delta\phi(x), \tag{14.51}$$

where $\delta\phi$ is considered to be small. The mean field is defined by the fact that the first-order variation of $E[\phi]$ should vanish:

$$0 = \delta E[\phi] = \int d^D x \left[ -\nabla^2 m(x) + 2r_0 m^2(x) + 4u_0 m^3(x) - h_0 \delta^D(x) \right] \delta\phi(x). \tag{14.52}$$

Since $\delta\phi$ is arbitrary, we must have

$$-\nabla^2 m(x) + 2r_0 m(x) + 4u_0 m^3(x) = h_0 \delta^D(x). \tag{14.53}$$

This is an inhomogenous nonlinear Schrödinger equation (NLSE), which occurs in such diverse fields as plasma physics, quantum optics, superfluidity, and, in a relativistic version, theory of elementary particles.

To get a solvable equation, we drop the nonlinear $m^3$ term, arguing that $m$ is small for $t > 0$. The equation becomes

$$-\nabla^2 m(x) + 2r_0 m(x) = h_0 \delta^D(x). \tag{14.54}$$

Taking the Fourier transform of both sides we obtain

$$\left( k^2 + 2r_0 \right) \tilde{m}(k) = h_0, \tag{14.55}$$

where $\tilde{m}(k)$ is the Fourier transform of $m(x)$ :

$$\tilde{m}(k) = \int d^D x\, e^{-ik \cdot x} m(x),$$

$$m(x) = \int \frac{d^D k}{(2\pi)^D} e^{ik \cdot x} \tilde{m}(k). \tag{14.56}$$

Thus

$$\tilde{m}(k) = \frac{h_0}{k^2 + 2r_0},$$
(14.57)

and the inverse transform gives

$$m(x) = h_0 \int \frac{d^D k}{(2\pi)^D} \frac{e^{ik \cdot x}}{k^2 + 2r_0}.$$
(14.58)

For large $|x|$ for $D > 2$ we have

$$m(x) \approx C_0 \frac{e^{-|x|/\xi}}{|x|^{D-2}},$$
(14.59)

where $C_0$ is a constant, and

$$\xi = (2r_0)^{-1/2} = (2bt)^{-1/2}$$
(14.60)

is called the *correlation length*. Its behavior at the critical point defines the critical exponent $v$:

$$\xi \sim t^{-v}.$$
(14.61)

According to our calculation

$$v = \tfrac{1}{2}.$$
(14.62)

This is a mean-field value, since we have ignored fluctuations.

The correlation length $\xi$ is finite at temperatures above the critical point, and sets a scale for the exponential decay of correlations with distance. At the critical point $\xi$ diverges. There is no length scale at this point, and the exponential law is replaced by a power law. According to (14.59), the power law has the form $|x|^{2-D}$, but this is not reliable, because we have neglected the $\phi^4$ term in the potential. A more careful analysis gives

$$m(x) \sim |x|^{2-D-\eta} \ (t = 0).$$
(14.63)

The dimension of space seems to have changed to $D+\eta$, which is a critical exponent called the "anomalous dimension".

## 14.8  Universality

We cannot resolve spatial structures smaller than $\xi$, because the field organizes itself into uniform blocks of approximately that size. As we approach the critical point, $\xi$ increases, and we lose resolution. At the critical point, when $\xi$ diverges, we cannot see any details at all. Only global properties, such as the dimension of space, or the number of degrees of freedom, distinguishes one system from another. That is why systems at the critical point fall into universality classes characterized by their critical exponents.

Imagine that you are blind-folded in a room, which you can probe only with a very long pole. You can find out whether the room is 1D, 2D or 3D by trying to move the pole, but you can learn little else. Similarly, if you put on very dark eyeglasses, you would conclude that all places on Earth are the same, characterized by a 24-hour light-dark cycle. To experience something new, you would have to go to Mars. (Or take those glasses off.)

## Problems

**14.1**   A highway system is proposed to connect four cities located at the corners of a unit square, as shown in the upper panel of the accompanying sketch. Show that the total length of the system can be reduced by adopting either of the schemes in the lower panel. Find the distance $a$ that minimizes the length.

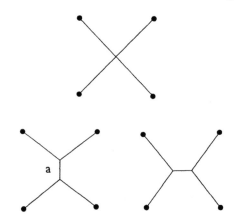

This is an example of broken symmetry. The invariance of the square under a 90° rotation is expressed through the fact that there are two minimal schemes that go into each other under the rotation.

**14.2**   In the mean-field treatment of the Ising model, find the magnetic moment $M$ from (14.12) at $h = 0$ in the neighborhood of the transition temperature $T_c$, where $M$ is small. Show $M \approx \sqrt{3t}$, hence $\beta = \frac{1}{2}$.

**14.3**   Check the fluctuation–dissipation theorem (14.48) in mean-field theory, by verifying that $\int d^D x \, m(x)$ and $\chi$ have the same critical exponent.

**14.4**   Verify (14.59) for $D = 3$, i.e.

$$m(\mathbf{x}) = \int \frac{d^3 k}{(2\pi)^3} \frac{e^{i\mathbf{k}\cdot\mathbf{x}}}{k^2 + 2r_0} = \frac{1}{4\pi |\mathbf{x}|} e^{-\sqrt{2r_0}|\mathbf{x}|}.$$

**14.5**   To describe a structural phase transition, for example the cubic-to-tetragonal transition of barium titanate ($BaTiO_3$), it is necessary to include a strain parameter $\varepsilon$ in the free energy as well as the order parameter $\eta$. Take the Landau free energy in the mean-field approximation to be

$$E(\eta, \varepsilon) = at\eta^2 + b\eta^4 + c\varepsilon^2 + g\eta^2\varepsilon,$$

where $t = T - T_c$, and $a, b, c, g$ are positive constants. In particular, $g$ is called the coupling constant.

(a) Minimize the free energy with respect to $\varepsilon$ for fixed $\eta$, to determine $\bar{\varepsilon}$ as a function of $\eta$.

(b) Obtain the effective free energy for $\eta$ alone. What is the new "renormalized" value, $\tilde{b}$, of $b$?

(c) What happens to $\tilde{b}$ as a function of $g$? What happens to the phase transition as a function of $g$?

**14.6**   The nematic liquid crystal used in displays can be described by an order parameter $S$ corresponding to the degree of alignment of molecular directions. In the ordinary fluid phase $S = 0$. The transition between the ordinary fluid phase and a nematic phase can be modeled via the mean-field Landau free energy

$$E(S) = atS^2 + bS^3 + cS^4$$

where $t = T - T_0$, and $a, b, c$ are positive constants. The third-order coefficient $b$ is usually small.

(a) Sketch $E(S)$ for a range of $T$, from $T \gg T_0$, through $T = T_0$ to $T < T_0$. Comment on the value of $S$ at the minimum of $E(S)$ at each value of $T$ considered.

(b) What are the conditions for $E(S)$ to be minimum?

(c) Find the transition temperature $T_c$. (It is not $T_0$.)

(d) Is there a latent heat associated with the phase transition? If so, what is it?

(e) How does the order parameter vary below $T_c$?

**14.7**   A system has a two-component order parameter: $\{\phi_1, \phi_2\}$. The mean-field Landau free energy is

$$E(\phi_1, \phi_2) = E_0 + at\left(\phi_1^2 + \phi_2^2\right) + b\left(\phi_1^2 + \phi_2^2\right)^2 + c\left(\phi_1^4 + \phi_2^4\right)$$

where $t = T - T_c$, and $a, b$ are positive constants.

(a) Represent the order parameter as a vector on a plane. Use polar coordinates to write $\phi_1 = \phi \cos\theta$, $\phi_1 = \phi \sin\theta$, where $\phi = \sqrt{\phi_1^2 + \phi_2^2}$. Minimize the free energy with respect to $\phi$, and show that its minimum occurs at the minimum of

$$\tilde{b} = b + c\left(\cos^4\theta + \sin^4\theta\right)$$

(b) Find the values of the order parameter in the ordered phase for the three possibilities $c < 0, c = 0, c > 0$.

(c) For $c < 0$, suppose the order parameter is parallel to the $x$-axis. Calculate the susceptibility $\chi_x$ with respect to an external field applied in the $x$-direction, and $\chi_y$ in the $y$-direction.

# Chapter 15

# Superfluidity

## 15.1  Condensate wave function

In the ideal Bose gas, a finite fraction of the particles occupy the same single-particle state below a critical temperature, forming a Bose–Einstein condensate. When the particles have mutual interactions, however, single-particle states are no longer meaningful. Nevertheless, we can still define a *condensate wave function* $\psi(\mathbf{r})$, as the quantum amplitude for removing a particle from the condensate at position $\mathbf{r}$. Its complex conjugate $\psi^*(\mathbf{r})$ is the amplitude for creating a particle.

We can imagine creating a particle in the condensate by inducing a transition into it. As we saw in Section 10.2, Bose enhancement makes the rate proportional to the existing density of bosons. Thus, the creation amplitude $\psi^*(\mathbf{r})$ should be proportional to the square root of the density. We define $|\psi(\mathbf{r}, t)|^2 \, \mathrm{d}^3 r$ as the number of condensate particles in the volume element $\mathrm{d}^3 r$, so that the total number of particles in the condensate is

$$\int \mathrm{d}^3 r \, |\psi(\mathbf{r})|^2 = N_0. \tag{15.1}$$

As we shall see, $\psi$ satisfies a nonlinear Schrödinger equation, and differs from an ordinary wave function in that the normalization cannot be changed arbitrarily.

We take $\psi$ to be the field whose ensemble average gives the order-parameter density for Bose–Einstein condensation. The Landau free energy is taken to be

$$E[\psi, \psi^*] = \int \mathrm{d}^3 r \left[ \frac{\hbar^2}{2m} |\nabla \psi|^2 + (U - \mu)\psi^* \psi + \frac{g}{2} \left( \psi^* \psi \right)^2 \right], \tag{15.2}$$

where $|\nabla \psi|^2 = \nabla \psi^* \cdot \nabla \psi$, $m$ is the mass of the particles, $U(\mathbf{r})$ is an external potential, and $\mu$ is the chemical potential that determines $N_0$. Because of the presence of the term $(\psi^* \psi)^2$, the normalization of $\psi$ cannot be changed without changing $g$.

In the Ginsburg–Landau philosophy, (15.2) is to be used near the critical temperature $T_c$, where $\psi \to 0$. On the other hand, (15.2) can be derived from a microscopic point of view (Huang 1987, Section 13.8), as an effective Hamiltonian for a dilute Bose gas at low temperatures, with

$$g = \frac{4\pi a \hbar^2}{m},$$  (15.3)

where $a$ is the "scattering length," the diameter of the hard-sphere equivalent of an atom. The conditions for the validity of this picture are

$$n^{1/3} a \ll 1,$$

$$\frac{a}{\lambda} \ll 1,$$  (15.4)

where $n = N_0/V$, and $\lambda = \sqrt{2\pi\hbar^2/mkT}$ is the thermal wavelength. Therefore, we can use (15.2) either in the neighborhood of $T = T_c$ or $T = 0$.

The grand partition function of the condensate is given by

$$Q = \int (D\psi)(D\psi^*) e^{-\beta E[\psi, \psi^*]},$$  (15.5)

where $\int (D\psi)(D\psi^*)$ denotes functional integrations over the real and imaginary parts of $\psi$ independently. Being a complex number, the order parameter $\psi$ has both a modulus and a phase. The Landau free energy is invariant under the phase change

$$\psi \rightarrow e^{i\chi} \psi,$$

$$\psi^* \rightarrow e^{-i\chi} \psi^*,$$  (15.6)

where $\chi$ is a constant. This symmetry is known as *global gauge invariance*, and implies the conservation of particles. It is spontaneously broken when the ensemble average

$$\langle \psi(x) \rangle = \frac{1}{Q} \int (D\psi)(D\psi^*) e^{-\beta E[\psi, \psi^*]} \psi(x)$$  (15.7)

does not vanish, since this quantity must have a definite phase. Just as spontaneous magnetization breaks rotational invariance, Bose–Einstein condensation breaks global gauge invariance.

## 15.2 Mean-field theory

We shall use the mean-field approximation, and take $\langle \psi \rangle$ to be the function that minimizes the Landau free energy, neglecting all fluctuations. From now on, we write $\psi$ in place of $\langle \psi \rangle$ to simplify the notation, and call it the order parameter.

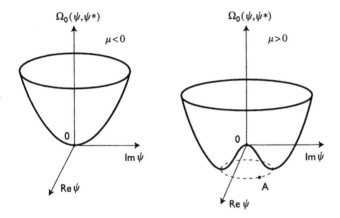

Figure 15.1 The potential plotted over the complex $\psi$-plane. For $\mu > 0$ the ground state corresponds to some point A, with a definite phase, thus breaking global gauge invariance. The Goldstone mode corresponds to A running along the dashed circle as the spatial location changes.

In the absence of an external field ($U = 0$), $\psi$ is uniform, and the kinetic term in the Landau free energy vanishes:

$$E[\psi, \psi^*] = V\Omega_0(\psi, \psi^*). \tag{15.8}$$

The potential

$$\Omega_0(\psi, \psi^*) = \frac{g}{2}|\psi|^4 - \mu|\psi|^2 \tag{15.9}$$

is plotted in Fig. 15.1 over the complex $\psi$-plane. For $\mu < 0$ the only minimum occurs at $\psi = 0$. For $\mu > 0$ the potential has a wine-bottle shape, with a continuous distribution of minima along a trough, shown as a dashed circle in Fig. 15.1.

In the ground state $\psi(x)$ is at the same point A for all space. Since it has a fixed phase, gauge invariance is spontaneously broken. It is re-expressed through the existence of the Goldstone mode, which corresponds to choosing a slightly different A for a slightly different $x$ (see Problem 15.9).

Carrying out the minimization of $\Omega_0$, we obtain

$$|\psi| = \begin{cases} 0 & (\mu < 0), \\ \sqrt{\mu/g} & (\mu > 0). \end{cases} \tag{15.10}$$

When $\mu < 0$, there is no condensate. When $\mu > 0$ the normalization condition (15.1) gives

$$\mu = \frac{gN_0}{V}. \tag{15.11}$$

This is the value of the chemical potential in the neighborhood of $T = 0$, where there is practically no thermal cloud, and $N_0$ is essentially the number of particles in the system. In the ideal Bose gas, we have $\mu = 0$ in the condensed phase. This is no longer true in this model, owing to the interparticle interactions represented by $g$. The chemical potential now depends on the condensate density, giving the condensate a finite compressibility.

We can use our model in the neighborhood of $T = T_c$, where $\mu$ changes sign. We assume as usual

$$\mu = bt, \tag{15.12}$$

where $t = 1 - (T/T_c)$. By comparison with (15.11), we obtain

$$\frac{N_0}{V} = \begin{cases} 0 & (t < 0), \\ bt/g & (t > 0). \end{cases} \tag{15.13}$$

In this approximation, of course, all the critical exponents are mean-field exponents.

## 15.3 Gross–Pitaevsky equation

From now on, we concentrate on the low-temperature aspect of the model. When there is a non-uniform external potential $U(\mathbf{r})$, the order parameter is no longer constant. Minimization of the Landau free energy (15.2) leads to a nonlinear Schrödinger equation (NLSE):

$$\left[ -\frac{\hbar^2}{2m} \nabla^2 + U + g\,|\psi|^2 \right] \psi = \mu\psi, \tag{15.14}$$

which is known as the *Gross–Pitaevsky equation* in the present context. This can be generalized to a time-dependent equation

$$\left[ -\frac{\hbar^2}{2m} \nabla^2 + U + g\,|\psi|^2 \right] \psi = i\hbar\frac{\partial\psi}{\partial t}, \tag{15.15}$$

which reduces to the previous equation if we put $\psi(\mathbf{r}, t) = \psi(\mathbf{r}, 0) \exp(-it\mu/\hbar)$. The normalization $\int d^3r\,|\psi|^2$ is a constant of the motion. The time-dependent equation not only describes the condensate, but also the Goldstone mode (see Problem 15.9).

Bose–Einstein condensation has been experimentally achieved in dilute gases of bosonic alkali atoms confined in an external potential (Ketterle 1999). The condensate typically contains the order of $10^6$ particles, with a spatial extension of order $10^{-2}$ cm. The average density is of order $10^{12}$ cm$^{-3}$, and the transition temperature is of order $10^{-7}$ K. Compared with liquid $^4$He, the density is smaller by ten orders of magnitude, and the transition temperature lower by seven orders of magnitude.

At finite temperatures, a condensate is in equilibrium with a thermal cloud of non-condensed particles. In a uniform system there is no spatial separation between condensed and non-condensed particles, and the Bose–Einstein condensation is a condensation in momentum space. In a non-uniform potential, however, the condensation happens in real space, because the condensate wave function is peaked about the lowest point in the potential, and stands above the thermal cloud. Near the center, where the gradient of the wave function is small, we can neglect the kinetic term in (15.14), and obtain

$$|\psi(\mathbf{r})|^2 \approx \frac{1}{g}[\mu - U(\mathbf{r})]. \tag{15.16}$$

This is called the "Thomas–Fermi approximation" after an approximation in atomic physics.

The spatial condensation is experimentally demonstrated in Fig. 15.2, in a gas of Na atoms confined in a harmonic optical potential formed by laser beams. The density profiles of the gas at successively lower temperatures show the onset of

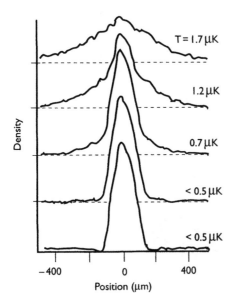

Figure 15.2 Density profile of a gas of Na atoms confined in a harmonic optical potential, at successively lower temperatures. Bose–Einstein condensation occurs at 1.7 μK, when a central peak emerges, representing the condensate wave function. The width of the peak is 0.03 cm, which is to be compared with a width of 0.005 cm of the harmonic-oscillator wave function in the absence of interparticle interactions. After Stenger (1998).

Bose–Einstein condensation at about $1.7 \times 10^{-6}$ K, when a peak representing the condensate wave function begins to emerge from the background. With decreasing temperature, the condensate grows at the expense of the thermal cloud. Below $0.5 \times 10^{-6}$ K almost all atoms are in the condensate. The condensate wave function fits (15.16), with a width of about 0.03 cm. This is to be compared with 0.005 cm, the width of the harmonic oscillator wave function for non-interacting atoms. We see that the repulsion between atoms broadens the peak considerably.

## 15.4   Quantum phase coherence

The order parameter embodies the idea of the macroscopic occupation of a single mode, and this implies quantum phase coherence. If you pull particles out of the condensate one by one, conceptually speaking, you would find that their wave functions all have the same phase. This coherence is the true essence of Bose–Einstein condensation, and can be demonstrated experimentally. The wave function of a condensate moving as a whole can be represented by a plane wave, except near the edges. Consider two separate condensates moving towards each other, and eventually overlapping, as illustrated in Fig. 15.3. The total wave function is the sum of the two wave functions, for an atom can be in either condensate:

$$\psi(r,t) = C_1 e^{i(\mathbf{k}_1 \cdot \mathbf{r} - \omega_1 t)} + C_2 e^{i(\mathbf{k}_2 \cdot \mathbf{r} - \omega_2 t)}, \tag{15.17}$$

where $C_i$ are constants, and $\hbar \omega_i = \hbar^2 k_i^2 / 2M_i$, where $M_i$ is the mass of the $i$th condensate ($i = 1, 2$). The density of the total system is given by

$$\begin{aligned}|\psi(r,t)|^2 &= \left| C_1 e^{i(\mathbf{k}_1 \cdot \mathbf{r} - \omega_1 t)} + C_2 e^{i(\mathbf{k}_2 \cdot \mathbf{r} - \omega_2 t)} \right|^2 \\ &= |C_1|^2 + |C_2|^2 + 2\mathrm{Re}\left[ C_1^* C_2 e^{i(\mathbf{k}_2 - \mathbf{k}_1) \cdot \mathbf{r} - i(\omega_2 - \omega_1)t} \right].\end{aligned} \tag{15.18}$$

The last term exhibits interference fringes. Fig. 15.4 shows a photograph of the fringes produced by interference between two Na condensates.

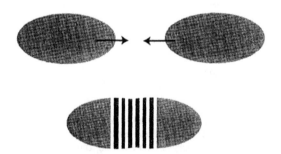

Figure 15.3 Two condensates approach each other, overlap, and exhibit interference fringes.

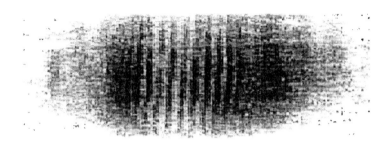

*Figure 15.4* Photograph of interference fringes produced by two overlapping condensates of Na atoms. The separation between fringes is $1.5 \times 10^{-3}$ cm. Courtesy W. Ketterle.

The ideal Bose gas has no phase coherence, unless we define it as the limit of an interacting gas. The ground state wave function of the ideal gas is a product of single-particle wave functions:

$$\Psi_0 = u_0(\mathbf{r}_1) \cdots u_0(\mathbf{r}_N). \tag{15.19}$$

We can change the phase of each factor $u_0$ independently, for that merely changes the overall phase factor, which is arbitrary. However, if there are interactions, no matter how weak, then the wave function acquires a correction:

$$\Psi_0 = [u_0(\mathbf{r}_1) \cdots u_0(\mathbf{r}_N)] + \Psi' \tag{15.20}$$

and we can no longer change the phase of the first factor relative to the second in an arbitrary manner. Thus, phase coherence relies on interactions, just as spontaneous magnetization relies on spin–spin attractions.

## 15.5 Superfluid flow

The time-dependent NLSE (15.15) conserves the number of particles in the condensate. We have a continuity equation

$$\frac{\partial n}{\partial t} + \nabla \cdot \mathbf{j} = \mathbf{0}, \tag{15.21}$$

where $n$ is the particle density, and $\mathbf{j}$ is the particle current density:

$$\mathbf{j}(\mathbf{r}) = \frac{\hbar}{2mi} \left( \psi^* \nabla \psi - \psi \nabla \psi \right). \tag{15.22}$$

Putting

$$\psi = \sqrt{n}e^{i\varphi}, \tag{15.23}$$

we have

$$\mathbf{j} = \frac{n\hbar}{m} \nabla \varphi, \tag{15.24}$$

from which we identify the *superfluid velocity*

$$\mathbf{v}_s = \frac{\hbar}{m} \nabla \varphi, \tag{15.25}$$

which describes a flow of the condensate without dissipation.

The circulation of the superfluid velocity field around a closed path $C$ is

$$\oint_C d\mathbf{s} \cdot \mathbf{v}_s = \frac{\hbar}{m} \oint_C d\mathbf{s} \cdot \nabla \varphi. \tag{15.26}$$

The integral on the right-hand side is the change of the phase angle upon traversing the loop $C$, and must be an integer multiple of $2\pi$, by continuity of the condensate wave function. Therefore the circulation is quantized:

$$\oint_C d\mathbf{s} \cdot \mathbf{v}_s = \frac{h\kappa}{m} \quad (\kappa = 0, \pm 1, \pm 2, \ldots). \tag{15.27}$$

A vortex is a flow pattern with a non-vanishing vorticity concentrated along a directed line called the vortex core. In Fig. 15.5 we show a vortex line, in which the vorticity is contained in a linear core, and a vortex ring, whose core is a closed curve. The circulation is zero around any closed loop not enclosing the vortex core, and non-zero otherwise, as illustrated by the loop $C$ in the figure.

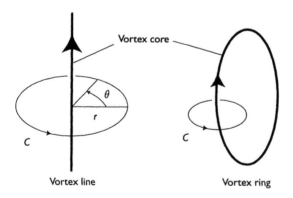

Figure 15.5 Vortex line and vortex ring.

For the vortex line, cylinderical symmetry dictates that the velocity field $v_s(r)$ depend only on the normal distance $r$ from the vortex core. Using (15.27), we have

$$\int_0^{2\pi} d\theta\, r v_s(r) = \frac{\hbar\kappa}{m}.\tag{15.28}$$

Thus

$$v_s(r) = \frac{\hbar\kappa}{mr},\tag{15.29}$$

which, according to (15.25), gives

$$\varphi = \kappa\theta,\tag{15.30}$$

where $\theta$ is the angle around the vortex core. This phase is defined only mod($2\pi$), but the superfluid velocity is unique. When $r \to 0$, the continuum picture breaks down at some atomic distance, which provides a cutoff to (15.29), and gives a finite radius to the vortex core.

## 15.6 Superconductivity

The Ginsburg–Landau theory was originally designed to describe superconductivity. Physically, the condensate is made up of *Cooper pairs,* which are two-electron bound states with spin zero. The order parameter satisfies an NLSE called the Ginsburg–Landau equation:

$$\left[ -\frac{\hbar^2}{2m}\left( \nabla - \frac{2ie\hbar}{c}\mathbf{A} \right)^2 + a\psi + b\,|\psi|^2 \right]\psi = 0,\tag{15.31}$$

where $a$ and $b$ are constants. This equation contains coupling to a vector potential $\mathbf{A}$, which gives an external magnetic field $\mathbf{B} = \nabla \times \mathbf{A}$. The coupling constant $e$ is the magnitude of the electron's charge. That is, the charge of the electron is $-e$ and the charge of a Cooper pair is $-2e$. This equation is invariant under the local gauge transformation

$$\mathbf{A} \to \mathbf{A} - \nabla\chi,$$
$$\psi \to \psi e^{-i(2e/\hbar c)\chi},\tag{15.32}$$

where $\chi(\mathbf{r})$ is an arbitrary function. We shall impose the Coulomb gauge condition $\nabla \cdot \mathbf{A} = 0$. The conserved current density is given by

$$\mathbf{j} = \frac{e\hbar}{mi}\left( \psi^*\nabla\psi - \psi\nabla\psi \right) - \frac{4e^2}{mc}|\psi|^2\,\mathbf{A}.\tag{15.33}$$

As we can verify by direct calculation, it satisfies $\nabla \cdot \mathbf{j} = 0$. We have included an overall factor of $2e$ by convention. The Ginsburg–Landau equation accounts for all the salient manifestations of superconductivity, including the Meissner effect and magnetic flux quantization (De Genne 1966, Chapter 6).

## 15.7 Meissner effect

In a uniform magnetic field, the order parameter is uniform, and the current density becomes

$$\mathbf{j} = -\frac{4e^2 n}{mc}\mathbf{A}, \tag{15.34}$$

where $n = |\psi|^2$ is the density of the condensate. This is known as the *London equation*. It is not explicitly gauge invariant, because we are using the Coulomb gauge. The vector potential must of course satisfy Maxwell's equation

$$\nabla \times \nabla \times \mathbf{A} = \frac{4\pi}{c}\mathbf{j}. \tag{15.35}$$

Using $\nabla \times \nabla \times \mathbf{A} = \nabla(\nabla \cdot \mathbf{A}) - \nabla^2 \mathbf{A}$ and $\nabla \cdot \mathbf{A} = 0$, we obtain

$$\left(\nabla^2 - \frac{16\pi e^2 n}{mc^2}\right)\mathbf{A} = 0. \tag{15.36}$$

For uniform $\mathbf{A}$, the solution is $\mathbf{A} = 0$. This means that the magnetic field must vanish inside a superconductor, and so should the current density. This is the *Meissner effect*.

If the superconductor is immersed in a uniform magnetic field, then the field can penetrate into the superconductor only within a layer of thickness called the *penetration depth*:

$$d = \sqrt{\frac{mc^2}{16\pi e^2 n}}. \tag{15.37}$$

A current density exists in this layer to shield the interior from the external magnetic field, and this makes up the superconducting current.

## 15.8 Magnetic flux quantum

Consider a hollow pipe made of a superconducting material like lead, with a total magnetic flux $\Phi$ inside the hollow pipe, as shown in Fig. 15.6. Inside the superconducting material the magnetic field is zero, and hence the vector potential has "pure gauge" form

$$\mathbf{A} = \nabla\chi. \tag{15.38}$$

On the other hand, its line integral along the closed loop $C$ must give the total flux:

$$\oint_C d\mathbf{s} \cdot \mathbf{A} = \Phi. \tag{15.39}$$

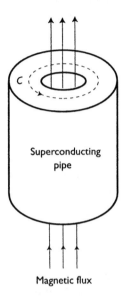

Superconducting
pipe

Magnetic flux

*Figure 15.6* Magnetic flux inside the superconducting pipe is quantized in units of $\Phi_0/2 = hc/2e$.

For a circular loop $C$, this can be satisfied by choosing

$$\chi = \frac{\theta\,\Phi}{2\pi},$$   (15.40)

where $\theta$ is an angle around the loop. However, the vector potential must be removable through a gauge transformation, for otherwise there will be a current inside the superconductor according to (15.34). In this gauge transformation, the order parameter undergoes the phase change

$$\psi \to \psi e^{-i(2e/\hbar c)\chi} = \psi e^{-i(2e\Phi/hc)\theta}.$$

Since $\psi$ must be periodic in $\theta$ with period 2, we have

$$\frac{2e\Phi}{hc} = \kappa,$$   (15.41)

where $\kappa$ is an integer. This quantization condition can be rewritten as:

$$\Phi = \frac{\kappa}{2}\Phi_0,$$   (15.42)

where

$$\Phi_0 = \frac{hc}{e} \approx 10^{-7}\,\mathrm{Gauss\,cm^2},$$   (15.43)

is the *magnetic flux quantum*. If the quantization condition is violated, then transient currents will be induced to bring the flux to a quantized value.

## 15.9  Josephson junction

Writing the order parameter in the form (15.23), we can rewrite the current density (15.33) as

$$\mathbf{j} = \frac{2e\hbar n}{m}\nabla\varphi - \frac{4e^2 n}{mc}\mathbf{A}. \tag{15.44}$$

The first term is a supercurrent density driven by a spatial variation of the phase $\varphi$. Consider two separate superconductors 1 and 2, with respective order parameters

$$\psi_1 = \sqrt{n_1}e^{i\varphi_1},$$
$$\psi_2 = \sqrt{n_2}e^{i\varphi_2}. \tag{15.45}$$

If $\varphi_1 > \varphi_2$, then current will flow from 1 to 2, provided there is an electrical link. We can provide a weak link by sandwiching between 1 and 2 a thin wafer of insulator, so that the Cooper pairs can pass through via quantum tunneling. This is called a *Josephson junction*. In a typical arrangement, the assembly is maintained at a temperature of 1.5 K. The sandwich has an area 0.025 × 0.065 cm², with an insulating layer of thickness 2000 Å.

As illustrated in Fig. 15.7, we apply a voltage difference $V$ between the super-conductors, so that a current $I$ flows across the junction. The reference point for

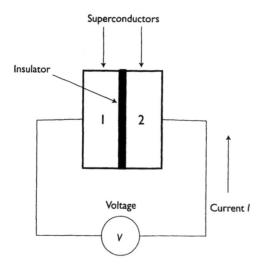

*Figure 15.7* A Josephson junction.

voltage is so chosen that the voltage is $-V/2$ at 1, and $V/2$ at 2. Since the tunneling link is very weak, we couple the two systems linearly. We model the equations for the order parameters after the NLSE, but neglect the nonlinear self-interactions:

$$i\hbar \frac{d\psi_1}{dt} = eV\psi_1 + K\psi_2,$$

$$i\hbar \frac{d\psi_2}{dt} = -eV\psi_2 + K\psi_1, \tag{15.46}$$

where $K$ is a coupling constant. The charge carriers are Cooper pairs with charge $-2e$. Our object is to study the current–voltage characteristic of the Josephson junction.

We now go to the representation (15.45). The first of the coupled equations becomes

$$\frac{i\hbar}{2\sqrt{n_1}} \frac{dn_1}{dt} - \hbar\sqrt{n_1}\dot{\varphi}_1 = eV\sqrt{n_1} + K\sqrt{n_2}e^{i(\varphi_2-\varphi_1)}, \tag{15.47}$$

where dots denote time derivatives. Equating real and imaginary parts on both sides, and putting

$$\varphi = \varphi_2 - \varphi_1, \tag{15.48}$$

we obtain the equations

$$\frac{dn_1}{dt} = \frac{2K}{\hbar}\sqrt{n_1 n_2}\sin\varphi,$$

$$\frac{d\varphi_1}{dt} = -\frac{eV}{\hbar} - \frac{K}{\hbar}\sqrt{\frac{n_2}{n_1}}\cos\varphi. \tag{15.49}$$

The second set of equations can be obtained by interchanging the indices 1 and 2, and reversing the sign of $V$:

$$\frac{dn_2}{dt} = -\frac{2K}{\hbar}\sqrt{n_1 n_2}\sin\varphi,$$

$$\frac{d\varphi_2}{dt} = \frac{eV}{\hbar} - \frac{K}{\hbar}\sqrt{\frac{n_1}{n_2}}\cos\varphi. \tag{15.50}$$

From these we obtain

$$\frac{d\varphi}{dt} = \frac{2eV}{\hbar} - \frac{K}{\hbar}\left(\sqrt{\frac{n_1}{n_2}} - \sqrt{\frac{n_2}{n_1}}\right)\cos\varphi. \tag{15.51}$$

Assuming $n_1 \approx n_2$, we neglect the second term on the right-hand side and get

$$\frac{d\varphi}{dt} = \frac{2eV}{\hbar}. \tag{15.52}$$

The Josephson current is defined by

$$I = e\frac{dn_1}{dt} = I_0 \sin\varphi, \tag{15.53}$$

where $I_0$ is a constant:

$$I_0 = \frac{2eK}{\hbar}\sqrt{n_1 n_2}. \tag{15.54}$$

The current–voltage characteristic is therefore given by

$$\begin{aligned} I &= I_0 \sin\varphi, \\ V &= \frac{\hbar}{2e}\frac{d\varphi}{dt}. \end{aligned} \tag{15.55}$$

As we shall see, the behavior of this system is most peculiar. If we apply a DC voltage, we get an AC current of such high frequency that it averages to zero. On the other hand, if we apply an AC voltage, we get a DC current.

### 15.9.1   DC Josephson effect

Let us apply a constant voltage

$$V = V_0. \tag{15.56}$$

We can integrate $d\varphi/dt = 2eV_0/\hbar$ to obtain

$$\varphi(t) = \varphi_0 + \frac{2eV_0 t}{\hbar}. \tag{15.57}$$

The current is given by

$$I(t) = I_0 \sin\left(\varphi_0 + \frac{2eV_0 t}{\hbar}\right). \tag{15.58}$$

For a typical voltage $V_0 = 10^{-3}$ V, we have

$$\frac{2eV_0}{\hbar} = 3.2 \times 10^{12} \text{ s}^{-1}. \tag{15.59}$$

Therefore the current oscillates with a very high frequency and averages to zero:

$$I_{av} = 0. \tag{15.60}$$

### 15.9.2 AC Josephson effect

Consider an AC voltage

$$V(t) = V_0 + V_1 \cos \omega t. \tag{15.61}$$

We can write

$$\frac{d\varphi}{dt} = \omega_0 + \omega_1 \cos \omega t, \tag{15.62}$$

where

$$\omega_0 = \frac{2e}{\hbar} V_0, \qquad \omega_1 = \frac{2e}{\hbar} V_1. \tag{15.63}$$

Thus

$$\varphi(t) = \varphi_0 + \omega_0 t + \frac{\omega_1}{\omega} \sin \omega t,$$

$$I(t) = I_0 \sin \left( \varphi_0 + \omega_0 t + \frac{\omega_1}{\omega} \sin \omega t \right). \tag{15.64}$$

To simplify the notation, put

$$A = \phi_0 + \omega_0 t, \qquad B = \frac{\omega_1}{\omega} \sin \omega t. \tag{15.65}$$

Suppose $\omega_1 \ll \omega_0$, then

$$
\begin{aligned}
I(t) &= I_0 \sin (A + B) = I_0 (\sin A \cos B + \sin B \cos A) \\
&\approx I_0 (\sin A + B \cos A) \\
&= I_0 \left[ \sin (\phi_0 + \omega_0 t) + \frac{\omega_1}{\omega} \sin \omega t \cos (\phi_0 + \omega_0 t) \right] \\
&= I_0 \sin (\phi_0 + \omega_0 t) \\
&\quad + I_0 \frac{\omega_1}{2\omega} [\sin (\phi_0 + (\omega + \omega_0)t) + \sin (\phi_0 + (\omega - \omega_0)t)].
\end{aligned}
\tag{15.66}
$$

Upon time-averaging, all oscillating terms go to zero, and only the last term survives:

$$I_{av} = \begin{cases} 0 & (\omega \neq \omega_0), \\ I_0 \sin \varphi_0 & (\omega \approx \omega_0). \end{cases} \tag{15.67}$$

The average current as a function of frequency has a sharp peak at

$$\omega_0 = \frac{2eV_0}{\hbar}. \tag{15.68}$$

Using the AC Josephson effect, the value of $e/h$ can be measured to very high accuracy.

## 15.10 The SQUID

Consider the presence of a vector potential $\mathbf{A}$, which gives rise to a magnetic field and electric field given by

$$\mathbf{B} = \nabla \times \mathbf{A},$$

$$\mathbf{E} = -\nabla V - \frac{1}{c}\frac{\partial \mathbf{A}}{\partial t}. \tag{15.69}$$

We generalize (15.55) by making the replacement

$$V \to V + \frac{1}{c}\frac{d}{dt}\int d\mathbf{r}\cdot\mathbf{A}, \tag{15.70}$$

where the integral is taken along some path in space. Thus, the voltage equation in (15.55) becomes

$$\frac{d\varphi}{dt} = \frac{2e}{\hbar}V + \frac{2e}{\hbar c}\frac{d}{dt}\int d\mathbf{r}\cdot\mathbf{A}. \tag{15.71}$$

Integrating this gives

$$\varphi(t) = \varphi_0 + \frac{2e}{\hbar}\int dt V + \frac{2e}{\hbar c}\int d\mathbf{r}\cdot\mathbf{A}. \tag{15.72}$$

The current is given by (15.55):

$$I = I_0 \sin\left[\varphi_0 + \frac{2e}{\hbar}\int dt\, V + \frac{2e}{\hbar c}\int d\mathbf{r}\cdot\mathbf{A}\right]. \tag{15.73}$$

All this is preparation for describing the SQUID (superconducting quantum interference device), a device to measure magnetic flux, with enough sensitivity to detect one flux quantum. The arrangement is shown in Fig. 15.8. Two Josephson junctions $a$ and $b$ are connected in parallel. The currents flowing through the junctions are respectively

$$I_a = I_0 \sin\left(\varphi_0 + \frac{2e}{\hbar c}\int_a d\mathbf{r}\cdot\mathbf{A}\right),$$

$$I_b = I_0 \sin\left(\varphi_0 + \frac{2e}{\hbar c}\int_b d\mathbf{r}\cdot\mathbf{A}\right). \tag{15.74}$$

If there is a magnetic field $\nabla \times \mathbf{A}$ through the loop, then

$$\int_a d\mathbf{r}\cdot\mathbf{A} - \int_b d\mathbf{r}\cdot\mathbf{A} = \oint d\mathbf{s}\cdot\mathbf{A} = \Phi, \tag{15.75}$$

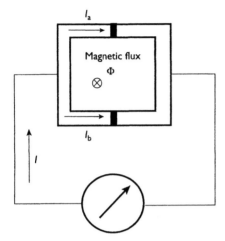

Figure 15.8 A SQUID consists of two Josephson junctions connected in parallel, forming a loop. Magnetic flux through the loop induces currents in the loop.

where $\Phi$ is the total magnetic flux going through the loop. We can put

$$
\int_a d\mathbf{r} \cdot \mathbf{A} = \frac{1}{2}\Phi + C_0,
$$
$$
\int_b d\mathbf{r} \cdot \mathbf{A} = -\frac{1}{2}\Phi + C_0,
$$

(15.76)

where $C_0$ is a constant that can be absorbed into $\varphi_0$. Thus

$$
I_a = I_0 \sin\left(\varphi_0 + \frac{e\Phi}{\hbar c}\right),
$$
$$
I_b = I_0 \sin\left(\varphi_0 - \frac{e\Phi}{\hbar c}\right).
$$

(15.77)

Adding these, we obtain the total current

$$
I = I_0\left[\sin\left(\varphi_0 + \frac{e\Phi}{\hbar c}\right) + \sin\left(\varphi_0 - \frac{e\Phi}{\hbar c}\right)\right]
$$
$$
= 2I_0 \sin\varphi_0 \cos\left(\frac{e\Phi}{\hbar c}\right) = 2I_0 \sin\varphi_0 \cos\left(\frac{2\pi\Phi}{\Phi_0}\right),
$$

(15.78)

where $\Phi_0$ is the elementary flux quantum given in (15.43). As $\Phi$ changes, the current oscillates with a period $\Phi_0$ as illustrated in Fig. 15.9.

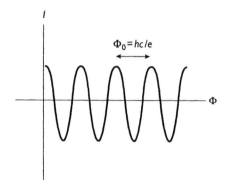

*Figure 15.9* Current response to magnetic flux in a SQUID. The oscillation period is one flux quantum. This gives the device great sensitivity.

## Problems

**15.1**   In the photograph in Fig. 15.4, the interference fringes between two Bose–Einstein condensates have a spacing of $1.5\times10^{-3}$ cm. Find the relative velocity between the two condensates.

**15.2**   Estimate the scattering length $a$ of sodium atoms from the data contained in Fig. 15.2, as follows. Use the Thomas–Fermi approximation (15.16) for the condensate profile. Use (15.11) for the chemical potential, with $N_0/V \approx 10^{11}$ cm$^{-3}$. The oscillator potential has a characteristic length $r_0 = \sqrt{\hbar/m\omega} = 5 \times 10^{-3}$ cm. The half width of the condensate is $3\times10^{-2}$ cm.

**15.3**   *Cold trapped atoms:* Consider a gas of $N$ non-interacting non-relativistic bosons of mass $m$ in an external harmonic-oscillator potential in 3D. The Hamiltonian of a particle is

$$H = \frac{p^2}{2m} + \frac{1}{2}m\omega^2 r^2,$$

where $p^2 = p_x^2 + p_y^2 + p_z^2$ and $r^2 = x^2 + y^2 + z^2$. Let $|\mathbf{n}\rangle$ be an eigenstate of $H$, where $\mathbf{n} = \{n_x, n_y, n_z\}$, with $n_x = 0, 1, 2, \ldots$, etc. The energy eigenvalues are

$$E_\mathbf{n} = \hbar\omega \left(n_x + n_y + n_z + \tfrac{3}{2}\right).$$

The fugacity $z$ is determined through

$$N = \sum_\mathbf{n} \frac{1}{z^{-1} \exp\left(E_\mathbf{n}/kT\right) - 1}.$$

The chemical potential is $\mu = kT \ln z$.

(a) Show that

$$\langle \mathbf{n}|r^2\,|\mathbf{n}\rangle = r_0^2\left(n_x + n_y + n_z + \tfrac{3}{2}\right),$$

where $r_0 = \sqrt{\hbar/m\omega}$. (*Hint:* Write $x^2$ in terms of creation and annihilation operators (see Problem 10.1).)

(b) Prove the virial theorem

$$\frac{1}{2m}\langle \mathbf{n}|\,p^2\,|\mathbf{n}\rangle = \frac{m\omega^2}{2}\langle \mathbf{n}|r^2\,|\mathbf{n}\rangle = \frac{1}{2}E_{\mathbf{n}}.$$

(c) Show that the thermal average of $r^2$ is

$$\langle r^2 \rangle = \frac{r_0^2}{N}\sum_{\mathbf{n}}\left(n_x + n_y + n_z + \tfrac{3}{2}\right)\frac{1}{z^{-1}\exp\left(E_{\mathbf{n}}/kT\right) - 1}.$$

**15.4**   Continuing with the previous problem, estimate the transition temperature of the Bose–Einstein condensation, as follows. For $kT \gg \hbar\omega$, it is a good approximation to replace the sum over the quantum numbers by an integral.

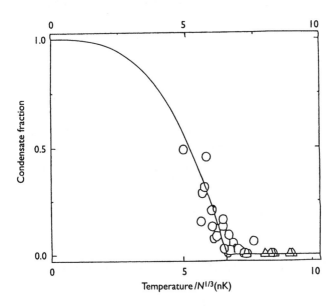

Calculate the temperature $T_0$ at which $\mu = \tfrac{3}{2}\hbar\omega$. Show that

$$kT_0 = b\omega N^{1/3},$$

where $b$ is a numerical constant. This is not intensive because of the external potential. The $N^{1/3}$ dependence is verified by experiments, as shown in the accompanying figure (after Mewes 1966).

**15.5** Continuing with the previous problem,

(a) Evaluate $\mu$ as a function of $N$ and $T$ in the classical limit, when the term $-1$ in the Bose distribution may be dropped. Show that

$$\mu = \frac{3}{2}\hbar\omega + 3kT \ln\left(\frac{T_0}{bT}\right).$$

(b) Evaluate the mean-square radius $\langle r^2 \rangle$ of the trapped gas in the classical limit. Show that

$$\langle r^2 \rangle = 3br_0^2 \left(\frac{T}{T_0}\right) N^{1/3}.$$

(c) Make qualitative plots of $\mu$ and $\langle r^2 \rangle$ as functions of $T$, extrapolate down to $T_0$, and comment on what happens below $T_0$.

**15.6** Consider a non-magnetic superconducting medium filling the half space $x > 0$. Suppose at $x = 0^-$ there is a uniform magnetic field normal to the $x$-axis. Find the magnetic field $B(x)$ in the medium.

**15.7** The state of a Josephson junction is specified by the superfluid phase difference $\varphi$ between two superconducting wafers, coupled through quantum-mechanical tunneling. The current–voltage characteristic is given through

$$I = I_0 \sin\varphi,$$
$$V = \frac{\hbar}{2e}\frac{d\varphi}{dt},$$

where $I$ is the current flowing through the junction and $V$ the voltage drop across the junction. The constant $I_0$ characterizes the coupling strength.

(a) Consider a circuit, in which a Josephson junction is connected in series with a resistor $R$, and a voltage source $U_0$. Set up the equation for the time development of $\varphi$.

(b) For $\kappa \equiv I_0 R/U_0 > 1$, show that the current approaches a limiting value $I_\infty$, while $V \to 0$

**15.8** *Soliton:* Show that the 1D NLSE has a soliton solution, as follows. From (15.15)

$$i\hbar\frac{\partial\psi}{\partial t} = -\frac{\hbar^2}{2m}\frac{\partial^2\psi}{\partial x^2} + g|\psi|^2\,\psi,$$

with $g = 4\pi a\hbar^2/m$. Rewrite the equation in the form

$$i\frac{\partial\psi}{\partial\tau} = -\frac{\partial^2\psi}{\partial x^2} + \lambda|\psi|^2\,\psi,$$

where $\tau = (\hbar/2m)\,t$, and $\lambda = 8\pi a$.

Show that the following is a solution for $\lambda > 0$:

$$\psi(x, \tau) = f(\theta)e^{-i(\sqrt{2b}\theta + 4b\tau)},$$

where $\theta = x - v\tau$, $b = v^2/8$, and

$$f(\theta) = b\sqrt{2/\lambda}\tanh(b\theta).$$

*Note*: The solution is a "dark soliton" propagating with velocity $v$. The case $\lambda < 0$ supports a "bright soliton". Both dark and bright solitons have been created in an optical fiber, which is governed by a NLSE.

**15.9** *Bogolubov spectrum:* The Goldstone mode in the condensate is the long-wavelength limit of a phonon mode described by the time-dependent NLSE (15.15).

(a) Take $U = 0$. Show that the uniform solution is $\psi_0 = \sqrt{n}\exp(-i\mu t/\hbar)$, where $n$ is the condensate density, and $\mu = gn$.
(b) Perturb the uniform solution by putting $\psi(x, t) = \psi_0 + f(x, t)\exp(-i\mu t/\hbar)$, where $f(x, t)$ is small, and to be kept only to first order. Show that

$$i\hbar\frac{\partial f}{\partial t} = -\frac{\hbar^2}{2m}\frac{\partial^2 f}{\partial x^2} + gn\left(f + f^*\right).$$

(c) Solve the equation by putting

$$f(x, t) = Ue^{-i(\omega - kx)} + V^*e^{i(\omega - kx)},$$

where $U$ and $V^*$ are complex constants. Show that

$$\omega = \frac{\hbar k}{2m}\sqrt{k^2 + 16\pi an}.$$

This is called the Bogolubov spectrum.
(d) What is the sound velocity?

# Chapter 16

# Noise

## 16.1 Thermal fluctuations

Thermodynamic quantities are supposed to be constant when the system is in thermal equilibrium. If we measure them with higher and higher precision, however, there will come a point when we will notice that their values undergo small fluctuations. For example, the pressure is the force per unit area exerted by atoms on a boundary surface, and it fluctuates because of the randomness of atomic impacts. The internal energy fluctuates because of energy exchange with a heat reservoir, which proceeds via atomic collisions. These fluctuations appear as thermal noise, and indicate the granular structure of matter.

We have calculated mean-square fluctuations in statistical mechanics, such as those for energy and density. We can usually ignore them, because they are vanishingly small in the thermodynamic limit. For example, according to (13.23), the mean-square fluctuation in the number of particles in a volume $V$ is

$$\langle N^2 \rangle - \langle N \rangle^2 = \frac{NkT}{V^2} \kappa_T, \tag{16.1}$$

where $n$ is the density and $\kappa_T$ is the isothermal compressibility. For an ideal gas this gives

$$\sqrt{\frac{\langle N^2 \rangle - \langle N \rangle^2}{\langle N \rangle^2}} = \frac{1}{\sqrt{N}}. \tag{16.2}$$

Numerically, this is utterly insignificant for a macroscopic volume of gas. However, in a volume of dimension 4000 Å, of the order of the wavelength of visible light, the number of atoms at STP is about $1.8 \times 10^6$, and the fractional rms fluctuation becomes 0.07%. This can be perceived indirectly through the scattering of light, as for example in the blue of the sky.

Whether thermal noise is significant, therefore, depends on the length scale we use. In this chapter, we study two types of noise accessible to direct observation, the Nyquist noise in an electrical resistor, and the Brownian motion of colloidal particles in suspension.

## 16.2  Nyquist noise

The thermal motion of electrons in metals produces electrical noise, which is audible when amplified, as in a radio signal. The spontaneous fluctuations in the voltage $V(t)$ or current $I(t)$ average to zero, but the rms fluctuations are not zero. In circuit elements that store energy, such as a capacitor $C$ or inductance $L$, these fluctuations can be obtained through the equipartition of energy (see Problem 6.14):

$$\tfrac{1}{2}C\overline{V^2} = \tfrac{1}{2}kT,$$

$$\tfrac{1}{2}L\overline{I^2} = \tfrac{1}{2}kT. \tag{16.3}$$

For a dissipative element such as a resistor, however, the fluctuations depend on its environment in a circuit.

For the spontaneous voltage fluctuation across the free ends of an open resistor, Nyquist (Nyquist 1928) derived the result

$$\overline{V^2} = 4RkT\Delta\nu, \tag{16.4}$$

where $R$ is the resistance of the resistor, $T$ the absolute temperature, and $\Delta\nu$ is the band width – the frequency range of the fluctuations. This result relates the voltage fluctuation to the resistance, and is an example of a fluctuation–dissipation theorem.

An intuitive argument for the result is as follows. The resistor at temperature $T$ exchanges energy with a heat reservoir, and the average heat dissipation in the frequency range $\Delta\nu$ is

$$\overline{I^2R} \propto kT\Delta\nu. \tag{16.5}$$

Using Ohm's law $I = V/R$, we obtain

$$\overline{V^2} \propto RkT\Delta\nu. \tag{16.6}$$

For quantitative results, consider a transmission line of length $L$ and impedance $R$. We terminate the transmission line at both ends with resistances $R$, so that traveling waves along the line are totally absorbed at the ends with no reflection. Let $V(t)$ be the voltage and $I(t)$ the current in one of the resistors. The frequency of the fundamental mode is

$$\nu_0 = \frac{c}{2L}. \tag{16.7}$$

At a finite temperature $T$ all higher modes are excited, with frequencies

$$\nu_n = n\nu_0, \quad n = 1, 2, 3, \ldots \tag{16.8}$$

The occupation number of the $n$th mode is $\left[\exp\left(\hbar\omega_n/kT\right) - 1\right]^{-1}$ where $\omega_n = 2\pi\nu_n$. Hence the energy residing in the $n$th mode is

$$E_n = \frac{\hbar\omega_n}{e^{\hbar\omega_n/kT} - 1}. \tag{16.9}$$

For $\hbar\omega_n/kT \ll 1$, we can use the approximation

$$E_n \approx kT. \tag{16.10}$$

In the band width $\Delta\nu$ there are $\Delta\nu/\nu_0$ modes, and the total energy is

$$E = \frac{kT\Delta\nu}{\nu_0} = \frac{2kTL}{c}\Delta\nu. \tag{16.11}$$

The energy can be regarded as residing in two traveling waves in opposite directions, and the time it takes to traverse the line is

$$t = \frac{L}{c}. \tag{16.12}$$

Thus the energy absorbed per second by each resistor is

$$W = \frac{E}{2t} = kT\Delta\nu \tag{16.13}$$

and the power delivered to each resistor is

$$\overline{I^2}R = kT\Delta\nu \tag{16.14}$$

As indicated in the lumped-circuit diagram of Fig. 16.1, the voltage across a resistor is $V = 2IR$. Therefore $I = V/2R$. Multiplying both sides by $I$, we obtain

$$I^2R = \frac{V^2}{4R}. \tag{16.15}$$

Hence

$$\overline{V^2} = 4kTR\Delta\nu. \tag{16.16}$$

This is known as the *Nyquist theorem*, and predicts a universal linear relation between $V^2$ and $R$, true for all materials. Numerically the voltage fluctuation is of

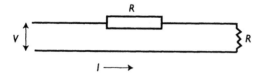

Figure 16.1 Lumped circuit diagram used in deriving the Nyquist theorem.

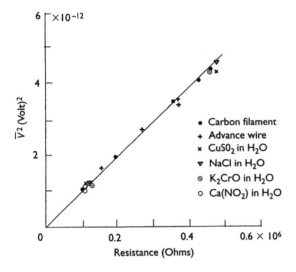

*Figure 16.2* Nyquist or Johnson noise: mean-square voltage fluctuation across the open ends of a resistor, as a function of temperature. The Nyquist theorem predicts the straight line, with a universal slope.

the order of microvolts, but Johnson (Johnson 1928) measured it, and verified the Nyquist relation, as shown in Fig. 16.2. He obtained Boltzmann's constant to an accuracy of 8%.

## 16.3  Brownian motion

In Brownian motion, we can see with our own eyes the manifestations of molecular thermal motion. This was unsettling to the nineteenth-century mind, which would rather stick to classical mechanics and thermodynamics. However, the experiments of Perrin on this subject, which won him the physics Nobel prize in 1926, demonstrated that matter is not the pristine continuum of classical thermodynamics, but made up of noisy atoms. In his words (Perrin 1909):

> When we consider a fluid mass in equilibrium, for example some water in a glass, all the parts of the mass appear completely motionless to us. If we put into it an object of greater density it falls and, if it is spherical, it falls exactly vertically. The fall, it is true, is the slower the smaller the object; but, so long as it is visible, it falls and always ends by reaching the bottom of the vessel. When at the bottom, as is well known, it does not tend again to rise, and this is one way of enunciating Carnot's principle (impossibility of perpetual motion of the second sort[1]).
>
> These familiar ideas, however, only hold good for the scale of size to which our organism is accustomed, and the simple use of the microscope suffices to

impress on us new ones which substitute a kinetic for the old static conception of the fluid state.

Indeed it would be difficult to examine for long preparations in a liquid medium without observing that all the particles situated in the liquid, instead of assuming a regular movement of fall or ascent, according to their density, are, on the contrary, animated with a perfectly irregular movement. They go and come, stop, start again, mount, descend, remount again, without in the least tending toward immobility. This is the Brownian movement, so named in memory of the naturalist Brown, who described it in 1827 (very shortly after the discovery of the achromatic objective), then proved that the movement was not due to living animalculae, and recognized that the particles in suspension are agitated the more briskly the smaller they are. . . .

The figure reproduced here (Fig. 16.3) shows three drawings obtained by tracing the segments which join the consecutive positions of the same granules of mastic[2] at intervals of 30 seconds. . . . They only give a very feeble idea of the prodigiously entangled character of the real trajectory. If the positions were indicated from second to second, each of these rectilinear segments would be replaced by a polygonal contour of 30 sides, relatively as complicated as the drawing here reproduced, and so on.

Over a vast range of the size of time steps, the Brownian path is a fractal, and a precise local velocity cannot be defined. In general, the path length $L$ and the step

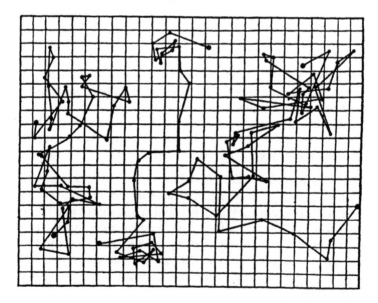

Figure 16.3 Brownian motion as sketched by Perrin, at thirty-second intervals. The grid size is 3.2 μ, and radius of the particle is 0.53 μ.

size $\tau$ are related through a power law:

$$L = a\tau^{1-D}, \tag{16.17}$$

where $a$ is a constant, and $D$ defines the *fractal dimension* of the path. A smooth curve has a length independent of the step size, and therefore corresponds to $D = 1$. It turns out that the Brownian path has $D = 2$. Thus, as we make the step size smaller and smaller, the path length increases inversely (that is, until we reach the molecular collision time).

## 16.4 Einstein's theory

Einstein thought of Brownian motion as a possible means to demonstrate the atomic nature of matter, without being aware of Brown's early observations. In his 1905 theory (Einstein 1905, 1906), a Brownian path is divided into finite steps of equal length, which eventually tends to zero. It is assumed that such a step, regardless of size, is the result of a large number of smaller steps. This is the idea of a "stochastic variable", a more complete discussion of which will be given in the next chapter.

Consider a particle moving in one dimension, with coordinate $x(t)$, which we know only through a distribution of probable values. During the time interval between $t$ and $t + \tau$, suppose $x$ undergoes the change

$$x \to x + \Delta \quad (-\infty < \Delta < \infty). \tag{16.18}$$

The probability distribution for $\Delta$ may be obtained as follows. Consider $N$ such particles, and let $dN$ be the number of particles displaced by a distance between $\Delta$ and $\Delta + d\Delta$, during the time interval between $t$ and $t + \tau$. The probability for the displacement to occur is given by the fraction of particles so displaced:

$$f_\tau(\Delta)\,d\Delta = \frac{dN}{N}, \tag{16.19}$$

which defines the probability per unit displacement $f_\tau(\Delta)$, a non-negative function with the properties

$$f_\tau(\Delta) = f_\tau(-\Delta),$$

$$\int_{-\infty}^{\infty} d\Delta\, f_\tau(\Delta) = 1. \tag{16.20}$$

The second condition means that $f_\tau(\Delta)$ should vanish faster than $\Delta^{-1}$ as $\Delta \to \infty$.

Let $n(x, t)$ be the density of particles. The number of particles being displaced out of the interval $dx$ during the time interval $\tau$ is given by

$$[n(x, t)\,dx]\left[\int_{-\infty}^{-dx} + \int_{dx}^{\infty}\right] d\Delta\, f_\tau(\Delta), \tag{16.21}$$

where the first factor is the number originally in $dx$, and the second factor is the probability of outflow. As $dx \to 0$, the gap between $-dx$ and $dx$ in the range of integration becomes insignificant, and the outflow is 100%. Therefore, all the particles found in $dx$ at time $t + \tau$ arrived during the time interval $\tau$. That is,

$$n(x, t + \tau)\, dx = \int_{-\infty}^{\infty} d\Delta f_\tau(-\Delta) n(x + \Delta, t)\, dx. \qquad (16.22)$$

Since $f_\tau(-\Delta) = f_\tau(\Delta)$ we have

$$n(x, t + \tau) = \int_{-\infty}^{\infty} d\Delta f_\tau(\Delta) n(x + \Delta, t). \qquad (16.23)$$

This is an expression of the conservation of particles.

We now expand $n(x, t + \tau)$ in powers of $\tau$, and $n(x + \Delta, t)$ in powers of $\Delta$:

$$n(x, t) + \tau \frac{\partial n(x, t)}{\partial t} + \cdots$$

$$= \int_{-\infty}^{\infty} d\Delta f_\tau(\Delta) \left[ n(x, t) + \Delta \frac{\partial n(x, t)}{\partial x} + \frac{\Delta^2}{2} \frac{\partial^2 n(x, t)}{\partial x^2} + \cdots \right]. \qquad (16.24)$$

The integral in the first term on the right is unity, and that in the second term vanishes because $f_\tau(\Delta)$ is an even function. Successive terms on the right-hand side should rapidly become smaller, as $\tau \to 0$. Thus, we have

$$\tau \frac{\partial n(x, t)}{\partial t} = \frac{\partial^2 n(x, t)}{\partial x^2} \frac{1}{2} \int_{-\infty}^{\infty} d\Delta\, \Delta^2 f_\tau(\Delta), \qquad (16.25)$$

which becomes exact in the limit $\tau \to 0$. Assuming the existence of the limit

$$D = \lim_{\tau \to 0} \frac{1}{2\tau} \int_{-\infty}^{\infty} d\Delta\, \Delta^2 f_\tau(\Delta), \qquad (16.26)$$

we obtain the diffusion equation

$$\frac{\partial n(x, t)}{\partial t} = D \frac{\partial^2 n(x, t)}{\partial x^2}. \qquad (16.27)$$

The diffusion constant $D$, which depends only on the second moment of the probability distribution $f_\tau(\Delta)$, can be taken from experiments.

## 16.5   Diffusion

The solution to the diffusion equation is

$$n(x,t) = \frac{1}{\sqrt{4\pi Dt}} e^{-x^2/(4Dt)}, \tag{16.28}$$

which is normalized such that

$$\int_{-\infty}^{\infty} dx\, n(x,t) = 1, \tag{16.29}$$

and $n(x,t)$ satisfies the initial condition

$$n(x,t) \underset{t\to 0}{\longrightarrow} \delta(x). \tag{16.30}$$

It gives the probability density of finding a particle at $x$ at time $t$, knowing that it was at $x = 0$ at $t = 0$. The generalization of the diffusion equation to 3D is

$$\frac{\partial n(\mathbf{r},t)}{\partial t} = D\nabla^2 n(\mathbf{r},t), \tag{16.31}$$

with solution

$$n(\mathbf{r},t) = \frac{1}{(4\pi Dt)^{3/2}} e^{-r^2/(4Dt)}. \tag{16.32}$$

To verify the diffusion law experimentally, Perrin translated each sketched Brownian path parallel to itself, so that they have a common origin in the plane of the paper. The theoretical distribution is thus

$$\frac{r\,dr}{(4\pi Dt)^{3/2}} \int_{-\infty}^{\infty} dz\, e^{-(r^2+z^2)/4Dt} = \frac{2r\,dr}{\rho^2} e^{-r^2/\rho^2}, \tag{16.33}$$

where $r$ is the radial distance in the plane, and $\rho = \sqrt{4Dt}$. For Perrin's experiments $\rho = 7.16\,\mu$. The distribution of 365 events is shown in Fig. 16.4, and the comparison with theory is shown in Fig. 16.5.

Einstein's simple and physical theory laid the foundation for a formal theory of stochastic processes. The important points are the following:

- The diffusion law is insensitive to the form of $f_\tau(\Delta)$. We can arrive at the same law by assuming that, at some small scale, a Brownian particle executes a random walk of equal step size (see Problem 5.9).
- The Brownian displacement, which is a sum of a large number of random steps, has a Gaussian distribution. This is the *central limit theorem*.

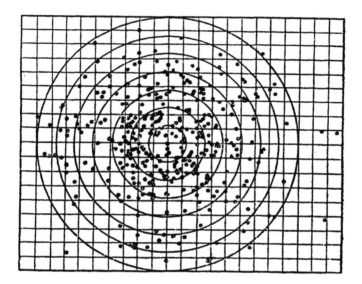

*Figure 16.4* Perrin translated 365 projected Brownian paths to a common origin, in order to check the diffusion law.

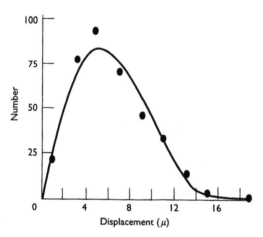

*Figure 16.5* Verification of the diffusion law in Brownian motion. Solid curve is theory and the dots represent data from Fig. 16.4.

## 16.6 Einstein's relation

The conservation of the number of Brownian particles is expressed by the continuity equation

$$\nabla \cdot \mathbf{j} + \frac{\partial n}{\partial t} = 0, \tag{16.34}$$

where $\mathbf{j}$ is the particle current density. Combining this with the diffusion equation $\partial n/\partial t = D\nabla^2 n$, we obtain $\nabla\cdot(\mathbf{j}+D\nabla n) = 0$, or

$$\mathbf{j} = -D\nabla n. \tag{16.35}$$

This shows that a density gradient produces a particle current (see Problem 16.8). Consider particles moving through a medium under an external force field

$$\mathbf{F}_{\text{ext}}(\mathbf{r}) = -\nabla U(\mathbf{r}). \tag{16.36}$$

The particles eventually reach a terminal drift velocity proportional to the force:

$$\mathbf{u} = \eta\mathbf{F}_{\text{ext}}, \tag{16.37}$$

which defines the mobility $\eta$. The drift current density $n\mathbf{u}$ produces a gradient in $n$, which in turn produces a counteracting diffusion current $\mathbf{j}$. In equilibrium these currents must balance each other, and hence

$$\mathbf{j}+n\mathbf{u} =0, \tag{16.38}$$

or

$$-D\nabla n - \eta n\nabla U = 0. \tag{16.39}$$

On the other hand, in equilibrium we must have the Boltzmann distribution

$$n(\mathbf{r}) = n(0)e^{-U(\mathbf{r})/kT}, \tag{16.40}$$

which gives

$$\nabla n = -\frac{n\nabla U(\mathbf{r})}{kT}. \tag{16.41}$$

Substituting this into (16.39), we obtain

$$\frac{Dn}{kT}\nabla U - \eta n\nabla U = 0, \tag{16.42}$$

and hence

$$D = kT\eta. \tag{16.43}$$

This is *Einstein's relation*, historically the first fluctuation–dissipation theorem. We shall return to this at the end of the chapter.

## 16.7  Molecular reality

At the end of the nineteenth century, three physical constants remained poorly known: Avogadro's number $A_0$, Boltzmann's constant $k$, and the fundamental charge $e$. Einstein's relation provides a way to experimentally measure Avogadro's number.

According to Stokes' law, a sphere moving at a terminal velocity $\mathbf{u}$ in a liquid experiences a frictional force

$$\mathbf{F} = 6\pi a\nu\mathbf{u}, \tag{16.44}$$

where $a$ is the radius of the sphere, and $\nu$ is the coefficient of viscosity of the liquid. The mobility is therefore

$$\eta = 6\pi a\nu. \tag{16.45}$$

By Einstein's relation we have

$$6\pi a\nu = \frac{D}{kT}. \tag{16.46}$$

Using the relation $k = R/A_0$, where $R$ is the gas constant and $A_0$ is Avogadro's number , we have

$$A_0 = \frac{6\pi a\nu RT}{D}. \tag{16.47}$$

Perrin obtained Avogadro's number from this relation, among others, and arrived at a weighted average:

$$A_0 = 7.05 \times 10^{23} \quad \text{(Modern value: } 6.02 \times 10^{23}). \tag{16.48}$$

Using this, Perrin was able to obtain $k$ and $e$ (see Problem 16.7).

Finally, we quote Perrin (Perrin 1909) on molecular reality and the second law of thermodynamics:

> It is clear that this agitation (the Brownian motion) is not contradictory to the principle of the conservation of energy... But it should be noticed, that it is not reconcilable with the rigid enunciations too frequently given to Carnot's principle[1] ... One must say: "On the scale of size which interests us practically, perpetual motion of the second sort is in general so insignificant that it would be absurd to take it into account."...
>
> On the other hand, the practical importance of Carnot's principle is not attacked, and I hardly need state at length that it would be imprudent to count upon the Brownian movement to lift the stones intended for the building of a house.....
>
> I think that it will henceforth be difficult to defend by rational arguments a hostile attitude to molecular hypotheses, which, one after another, carry

conviction, and to which at least as much confidence will be accorded as to the principles of energetics. As is well understood, there is no need to oppose these two great principles, the one against the other, and the union of Atomistics and Energetics will perpetuate their dual triumph.

## 16.8  Fluctuation and dissipation

The atomicity of matter not only gives rise to fluctuations of thermodynamic variables, it also explains why there is friction. As Einstein's relation $D/kT = \eta$ shows, fluctuations and dissipations are different aspects of the same physics:

- *Fluctuation*: In the absence of external force, the path of a Brownian particle fluctuates because of random molecular impacts. This is manifested through diffusion, characterized by the diffusion constant $D$.
- *Dissipation*: An external force is needed to drag a Brownian particle through the medium, because there is friction created by random molecular impacts, characterized by the mobility $\eta$.

These two aspects are illustrated in Fig. 16.6. The general pattern of the fluctuation–dissipation theorem is best exemplified by that from Ginsburg–Landau theory in Section 14.6:

$$\beta \int d^D x \, \langle \phi(x)\phi(0) \rangle = \chi, \tag{16.49}$$

where we can regard $\phi(x)$ as the magnetic moment density, and $\beta = 1/kT$. The integrand on the left-hand side is a correlation function in the absence of external magnetic field. The susceptibility on the right-hand side is defined by the linear response $\langle \phi \rangle_h = \chi h$, where $\langle \phi \rangle_h$ is the average magnetic moment density in the presence of a small uniform external magnetic field $h$.

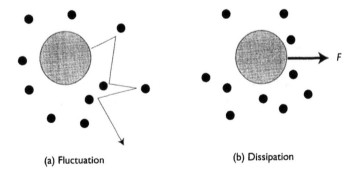

(a) Fluctuation                    (b) Dissipation

Figure 16.6  Two aspects of the motion of a body in a grainy medium: fluctuation (diffusion) and dissipation (mobility). They are united through Einstein's relation.

Table 16.1  Various forms of the fluctuation–dissipation
theorem

| Quantity | Fluctuation–dissipation theorem | Reference |
|---|---|---|
| Voltage | $\beta\left(\overline{V^2} - \bar{V}^2\right) = 4R\Delta\nu$ | Section 16.2 |
| Energy | $\beta\left(\overline{E^2} - \bar{E}^2\right) = TC_V$ | Section 12.4 |
| Density | $\beta\left(\overline{n^2} - \bar{n}^2\right) = V^{-1}\bar{n}^2\kappa_T$ | Section 13.4 |

In the case of Brownian motion, a more explicit form of the Einstein relation, derived later in Section 18.8, fits the pattern:

$$\beta \int_0^\infty dt \langle v(t)v(0)\rangle = \eta. \tag{16.50}$$

On the left-hand side we have the velocity correlation function in the absence of an external field, and on the right-hand side we have the susceptibility $\eta$ from the linear response $\langle v\rangle_F = \eta F$, where $\langle v\rangle_F$ is the average velocity in the presence of the external force $F$. In Table 16.1, we list the other fluctuation–dissipation theorems encountered in our studies.

It should be emphasized that these relations are valid only if

- the fluctuations represent small deviations from the state of thermodynamic equilibrium;
- the system responds in a linear manner to a small disturbance.

In view of these conditions, we would not be surprised if the fluctuation–dissipation theorem fails in highly nonlinear systems, or in systems that take years to reach thermal equilibrium, such as glasses.

## Problems

**16.1**  The Nyquist theorem holds at high temperatures. Find the first quantum correction to it. At what temperature would it be necessary to take this correction into account, for a resistor of length 1 mm?

**16.2**  Model a Brownian displacement as a path of fractal dimension 2. Start with a straight line of unit length between two fixed points. Draw a new path connecting the same endpoints with twice the length, and repeat this procedure as often as you can manage to do.

**16.3**  Show that Brownian particles in suspension form an ideal gas, with equation of state $PV = NkT$, where $P$ is the partial pressure of the suspended particles.

**16.4**  Verify the solution (16.28) to the diffusion equation, in particular, the normalization (16.29) and the initial condition (16.30).

**16.5**    (a) Perrin cited $\rho \equiv \sqrt{4Dt} = 7\cdot16\,\mu$, and $t = 30\,s$. Find the diffusion constant $D$ for the Brownian particles, and compare it with that for the self-diffusion constant of $O_2$ molecules at STP, which can be estimated from Section 7.5. How far would an $O_2$ molecule travel in the same time?

(b) Find the mobility of a Brownian particle at a temperature $T = 300\,K$. What force is required to drag the particle at a velocity of 1 cm/s?

**16.6**    Calculate the Boltzmann's constant $k = R/A_0$, and the electronic charge $e = F/A_0$, using Perrin's value of Avogadro's number $A_0 = 7.05 \times 10^{23}$. Here, $R = 8.32 \times 10^7\,erg\,K^{-1}$ is the gas constant and $F = 2.9 \times 10^{14}$ esu is the Faraday, the amount of charge dissociated in 1 g mol of substance through electrolysis.

**16.7**    The diffusion equation can be derived in many different ways. Apart from Einstein's derivation discussed in this chapter (which was historically one of the first) it can also be derived from the random walk, as shown in Problem 5.9. Here, we give another derivation, and some generalizations.

(a)  Start with (16.35), $\mathbf{j} = -D\nabla n$, as an empirical "constitutive relation" between the diffusion current $\mathbf{j}$ and the gradient of the density. Combining it with the continuity equation, obtain the diffusion equation in the absence of external force

$$\frac{\partial n}{\partial t} = D\nabla^2 n.$$

(b)  When there is an external force $\mathbf{F}_{ext}$, the total particle current is the sum of the diffusion current $-D\nabla n$ and the drift current $n\mathbf{F}_{ext}/\eta$, where $\eta$ is the mobility. Show that the diffusion equation is generalized to

$$\frac{\partial n}{\partial t} = D\nabla^2 n - \eta \mathbf{F}_{ext} \cdot \nabla n.$$

This is a special case of the Fokker–Planck equation.

(c)  If the diffusing particle can be absorbed by the medium with an absorption probability per second $V(\mathbf{r})$, show that the diffusion equation is generalized to

$$\frac{\partial n}{\partial t} = D\nabla^2 n - Vn.$$

In pure-imaginary time, this becomes the Schrödinger equation, with $D = \hbar/2m$.

# Chapter 17

# Stochastic processes

## 17.1 Randomness and probability

In physics, the term stochastic[1] refers to probabilistic considerations and the notion of probability is based on the frequency of occurrence of random events. From a physical point of view, therefore, the central questions are

- What is meant by randomness?
- How can we decide whether a process is random?

According to our intuitive notions, a random event is one whose outcome is uncertain and unpredictable. More specifically, it is an event for which an infinitesimal change in the initial conditions could radically affect the outcome.

The flipping of a "true" coin has 50% chance of being heads or tails. That is to say, the outcome is random. Actually, this defines what we mean by a true coin. If we toss a particular coin a million times, and heads turned up 49.9% of the time, we would conclude that either the coin is biased, or we have not tossed it enough times. For us to revise our notion of a true coin will require the shock of monumental failures. From this point of view, probability is a physical concept no different from any other. (For general reference on probability see Feller 1968.)

We now formulate these ideas more precisely. Imagine that $n$ independent experiments are performed under identical conditions. If outcome $A$ is obtained in $n_A$ of the experiments, then the probability that $A$ occurs is

$$P_A = \lim_{n \to \infty} \frac{n_A}{n}. \tag{17.1}$$

The following basic properties of probability follows from this definition:

- If two events A and B are mutually exclusive, then the probability that either $A$ or $B$ occurs is the sum of their probabilities $P_A + P_B$.
- If two events $A$ and $B$ are independent of each other, then the probability for their simultaneous occurrence is the product of their probabilities $P_A P_B$.

In an experiment whose outcome must be one of a number of mutually exclusive events labeled $1, \ldots, K$, the probability $P_i$ associated with the $i$th event is a real number satisfying

$$P_i \geq 0,$$

$$\sum_{i=1}^{K} P_i = 1. \tag{17.2}$$

There are practical difficulties in using this definition to experimentally measure probability. First, we do not have infinite time at our disposal; secondly no two actual experiments can be completely identical. Thus, the true function of this definition is to guide us in assigning a priori probabilities to events. Whether or not the assignment is appropriate is determined by confronting our theory with reality.

A *stochastic variable* is a quantity whose possible values occur with a certain probability distribution. It is defined when we give

- the range of its possible values $y$;
- the probability $P(y)$ for the occurrence of $y$.

The sum of two stochastic variables is a stochastic variable.

The simplest stochastic variable is the outcome $y$ for tossing a coin. Suppose the coin is biased, so that the probability for heads is $p$ and that for tails $(1 - p)$. The possible values may be taken to be

$$y = \begin{cases} 1 & \text{(heads)}, \\ 0 & \text{(tails)}, \end{cases} \tag{17.3}$$

with probability

$$P(1) = p,$$
$$P(0) = 1 - p. \tag{17.4}$$

In Einstein's theory of Brownian motion discussed in the last chapter, the displacement $\Delta$ over a time interval $\tau$ is a continuous stochastic variable, with range $-\infty < \Delta < \infty$. The probability that it has a value between $\Delta$ and $\Delta + d\Delta$ is $f_\tau(\Delta)d\Delta$. In this theory in the limit $\tau \to 0$, only the second moment of $f_\tau(\Delta)$ is relevant.

## 17.2  Binomial distribution

To put our notion of stochastic variables to use, we derive some probability distributions that can serve as tests for randomness.

Consider first the sum of a number of stochastic variables, for example, the result of tossing a coin $n$ times independently:

$$k = y_1 + \cdots + y_n. \tag{17.5}$$

The possible values are

$$k = 0, 1, \ldots, n. \tag{17.6}$$

The probability $P(k)$ can be obtained through the binomial theorem, which states that, in the expansion of $(1+x)^n$ in powers of $x$, the coefficient of $x^k$ is the binomial coefficient

$$\binom{n}{k} = \frac{n!}{k!(n-k)!}. \tag{17.7}$$

Let us write

$$(1+x)^n = \underbrace{(1+x)\cdots(1+x)}_{n \text{ factors}}. \tag{17.8}$$

To obtain $x^k$, we pick one $x$ each from $k$ of the factors on the right-hand side. The coefficient of $x^k$ is the number of ways we can choose $k$ factors out of the $n$ factors. Therefore, $\binom{n}{k}$ is the number of ways to choose $k$ things out of $n$ things.

Now, imagine $n$ identical coins being tossed simultaneously. There are $\binom{n}{k}$ ways of choosing the $k$ coins that turn up heads. The probability for each of the choices is $p^k(1-p)^{n-k}$, where $p$ is the probability for heads. The probability $P(k)$, which we redesignate as $B(k; n, p)$, is thus given by

$$B(k; n, p) = \frac{n!}{k!(n-k)!} p^k (1-p)^{n-k}. \tag{17.9}$$

This is called the *binomial distribution*, the probability for getting $k$ heads in $n$ tosses of a coin, with $p$ the intrinsic probability for heads. For a true coin $p = 1/2$.

The first few moments of the distribution are

$$\sum_{k=0}^{n} B(k; n, p) = 1,$$

$$\langle k \rangle = \sum_{k=0}^{n} k B(k; n, p) = np, \tag{17.10}$$

$$\langle k^2 \rangle = \sum_{k=0}^{n} k^2 B(k; n, p) = n^2 p^2 + np(1-p).$$

The variance, or mean-square fluctuation, is

$$\langle k^2 \rangle - \langle k \rangle^2 = np(1-p). \tag{17.11}$$

A plot of the distribution is shown in Fig. 17.1.

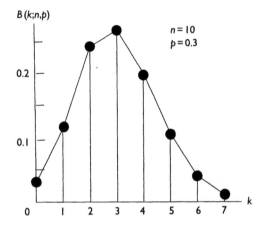

Figure 17.1 The binomial distribution gives the probability of getting $k$ heads in $n$ tosses of a biased coin, with $p$ the intrinsic probability for heads.

## 17.3   Poisson distribution

The *Poisson distribution* is the limit of the binomial distribution when $p \to 0$ and $n \to \infty$, with fixed $np = \sigma$:

$$P(k;\sigma) = \lim_{n \to \infty} B\left(k;n,\frac{\sigma}{n}\right). \qquad (17.12)$$

It is known as the "law of small probabilities" for this reason. With the help of Stirling's formula, it is straightforward to obtain

$$P(k;\sigma) = \frac{\sigma^k}{k!}e^{-\sigma}. \qquad (17.13)$$

The first few moments are

$$\sum_{k=0}^{\infty} P(k;\sigma) = 1,$$

$$\langle k \rangle = \sum_{k=0}^{\infty} kP(k;\sigma) = \sigma, \qquad (17.14)$$

$$\langle k^2 \rangle = \sum_{k=0}^{\infty} k^2 P(k;\sigma) = \sigma(\sigma+1).$$

Thus the variance is

$$\langle k^2 \rangle - \langle k \rangle^2 = \sigma. \qquad (17.15)$$

A 3D representation of the Poisson distribution is shown in Fig. 17.2.

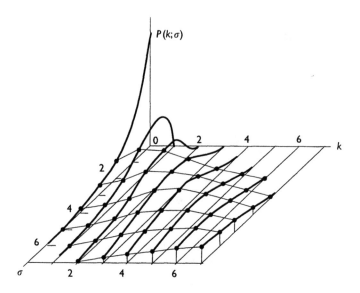

*Figure 17.2* The Poisson distribution (17.13). The variable $k$ is discrete, while $\sigma$ is continuous. Curves at constant large $\sigma$ approach a Gaussian distribution in $k$.

As an example of the Poisson distribution, suppose you try to sell walrus whiskers on the telephone. The demand for the product is essentially zero, but you make a thousand phone calls. What is the chance that you will make ten sales? Twenty sales? If you know the probability of a sale, you can calculate the chances via the Poisson distribution.

If you do not know the probability, and want to find out, you can make a thousand phone calls everyday for a year. At the end of the period you tally up the number of days $N(k)$ in which you made $k = 0, 1, 2, \ldots$ sales. If there is enough data, the histogram of $N(k)$ should resemble a Poisson distribution. You can then normalize it so the total area is 1, and try to fit it with a Poisson distribution by adjusting $\sigma$. The probability of a sale is then $p = \sigma/1000$.

## 17.4  Gaussian distribution

The Poisson distribution gives the probability for discrete counts $k$. It is peaked about $k = \sigma$, with width $\sqrt{\sigma}$. For sufficiently large $\sigma$, the values of $k$ in the neighborhood of the peak can be treated as continuous, and the Poisson distribution goes over to the *Gaussian distribution*, also known as the *normal distribution*.

In the neighborhood of $k = \sigma$, for $\sigma \gg 1$, we can use the Stirling formula to write

$$P(k;\sigma) \approx \frac{1}{\sqrt{2\pi k}} e^{k-\sigma+k\ln(\sigma/k)}. \tag{17.16}$$

Using the approximation

$$\ln\left(\frac{\sigma}{k}\right) = \ln\left[1 + \left(\frac{\sigma}{k} - 1\right)\right]$$
$$\approx \left(\frac{\sigma}{k} - 1\right) - \frac{1}{2}\left(\frac{\sigma}{k} - 1\right)^2, \tag{17.17}$$

we obtain

$$P(k;\sigma) \approx \frac{1}{\sqrt{2\pi k}} e^{-(k-\sigma)^2/2k}. \tag{17.18}$$

Since $k \approx \sigma$, we can replace $k$ by $\sigma$ everywhere, except in the quantity $(k - \sigma)^2$ in the exponent. Putting $x = k - \sigma$, we obtain the result

$$G(x;\sigma) = \frac{1}{\sqrt{2\pi\sigma}} e^{-x^2/2\sigma}, \tag{17.19}$$

where $x$ is regarded as a continuous variable. The moments are

$$\int_{-\infty}^{\infty} dx\, G(k;\sigma) = 1$$
$$\langle k \rangle = \int_{-\infty}^{\infty} dx\, x\, G(k;\sigma) = 0, \tag{17.20}$$
$$\langle k^2 \rangle = \int_{-\infty}^{\infty} dx\, x^2 G(k;\sigma) = \sigma.$$

Clearly, the variance is $\sigma$. Plots of the Gaussian distribution for different $\sigma$, are shown in Fig. 17.3.

## 17.5  Central limit theorem

We often work with the sum of a large number of stochastic variables. For example, to test the "trueness" of a coin, we make a large number $n$ of tosses, with outcomes $y_i$ ($i = 1,\ldots,n$), and consider the probability distribution of $k = \sum_i y_i$. For sufficiently large $n$, we should have the Gaussian distribution (17.19), which is centered at $\sigma = np$, with width $\sqrt{\sigma}$. Changing the variable to

$$z = \frac{1}{n}\sum_{i=1}^{n} y_i, \tag{17.21}$$

we obtain the Gaussian distribution

$$\frac{1}{\sqrt{2\pi\sigma}} e^{-(z-p)^2/2\sigma}. \tag{17.22}$$

The coin is true if $p = \frac{1}{2}$.

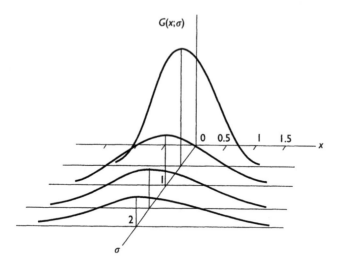

Figure 17.3 The Gaussian distribution (17.19). The central limit theorem states that the sum of a large number of stochastic variables obeys this distribution, independent of the probability distributions of the summands.

As another example, let $y_1, y_2, \ldots$ represent the displacements in the coordinate of a Brownian particle at successive time intervals $\tau \approx 10^{-10}$ s, on a molecular scale. If we make observations at one-second intervals, we are measuring the sum

$$x = \Delta_1 + \Delta_2 + \cdots + \Delta_n, \tag{17.23}$$

where $n$ is of the order $10^{10}$. We learned in the last chapter that, regardless of the detailed form of the probability $P(\Delta)$, the distribution of $x$ is a Gaussian

$$\frac{1}{\sqrt{4\pi Dt}} e^{-x^2/(4Dt)}, \tag{17.24}$$

where $D$ is proportional to the second moment of $P(\Delta)$, and $t$ is the observation time in seconds.

Summarizing these results, we can assert: *The sum of a large number of stochastic variables obeys the Gaussian distribution, regardless of the probability distributions of the individual stochastic variables.*

This is the *central limit theorem*. The important property of the Gaussian distribution is its universality, which makes its "bell curve" so ubiquitous, from error analysis in laboratory experiments to the forecast of longevity in a population.

## 17.6  Shot noise

Shot noise consists of a series of events randomly distributed over a long period of time. Practical examples abound:

- raindrops impinging on a window pane during a rainstorm,
- electrons arriving at the anode of a vacuum tube,
- cars crossing an intersection during rush hour,
- customers going up to a service counter.

In reality, of course, these events are subject to modulations. But, during a stretch of time when they appear to be in a steady state, we model them as a random stream.

We can map this problem into the coin-tossing problem discussed earlier. Let us divide the time axis into bins, each of duration $T$, and let $K_i$ be the number of events in the $i$th bin, as shown in Fig. 17.4. We regard the collection of bins as a statistical ensemble describing the process. The average frequency of events is defined by

$$\nu = \lim_{M \to \infty} \frac{K_1 + \cdots + K_M}{MT}. \tag{17.25}$$

Let the unit of time be $\Delta t$. The probability of a hit during $\Delta t$ is $p = \nu \Delta t$. We consider $\Delta t \to 0$, so $p$ becomes vanishingly small. Then, getting a hit is like getting heads in tossing a biased coin, with intrinsic probability $p$. There are $n = T/\Delta t$ intervals in the bin, and we can think of them as $n$ tosses of the coin. Thus, the probability of getting $k$ hits in $n$ tries is given by the binomial distribution $B(k; n, p)$. Now going to the limit $n \to \infty$, $p \to 0$, with fixed $np = \sigma$, where $\sigma$ is given by

$$\sigma = \nu T, \tag{17.26}$$

the probability of getting $k$ hits in time $T$ is given by the Poisson distribution

$$P(k; \sigma) = \frac{\sigma^k}{k!} e^{-\sigma}. \tag{17.27}$$

The average number of hits is $\langle k \rangle = \sigma$. The distribution depends on $T$ only through $\sigma$.

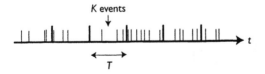

Figure 17.4 Shot noise consists of a stream of events distributed randomly over a long period of time. Here, the time axis is divided into periods of equal durations, and the collection of these periods form a statistical ensemble.

Suppose an event happening at time $t = 0$ produces a measurable effect $f(t)$, such as the sound of a rain drop, or a current triggered by an electron. The output from the streams of random events is represented by the function

$$I(t) = \sum_{k=-\infty}^{\infty} f(t - t_k). \tag{17.28}$$

Generally, $f(t)$ makes a "plop" that is zero before $t = 0$, and is significant only for a finite time $\Delta$. Thus, at any given $t$, only a finite number of events contribute significantly to the output function, namely, those that lie within the width $\Delta$. As an illustration, let us represent the sound of a raindrop arriving at time $t = 0$ by

$$f(t) = \begin{cases} 0 & (t < 0), \\ e^{-\lambda t} & (t > 0), \end{cases} \tag{17.29}$$

where $\lambda$ is a constant. Then $I(t)$ has the form shown in Fig. 17.5.

*Campbell's theorem* states the simple results

$$\langle I(t) \rangle = \nu \int_{-\infty}^{\infty} dt f(t),$$
$$\langle I^2(t) \rangle - \langle I(t) \rangle^2 = \nu \int_{-\infty}^{\infty} dt f^2(t), \tag{17.30}$$

where $\langle I(t) \rangle$ denotes the average over an ensemble of time periods of duration $T$, with $t$ held fixed in that period. The average turns out to be independent of $t$, in the limit $T \to \infty$.

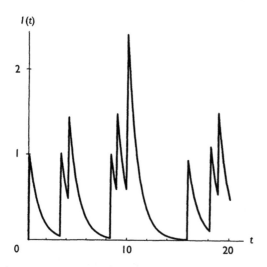

Figure 17.5 The sound of raindrops. A drop arriving at $t = 0$ produces sound represented by $f(t) = \theta(t) e^{-t}$.

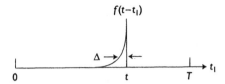

$f(t-t_1)$

Figure 17.6 The integral of the function over $t_1$ is the area under the curve, except when $t$ is within a width $\Delta$ of the boundaries.

To prove the first statement of Campbell's theorem, we first consider the cases where exactly $k$ events arrive in the period $T$, and average over the arrival times $t_1, \ldots, t_k$. We then average over the Poisson distribution for $k$:

$$\langle I(t) \rangle = \sum_k P(k; \sigma) \int_0^T \frac{dt_1}{T} \cdots \int_0^T \frac{dt_k}{T} \left[ f(t - t_1) + \cdots f(t - t_k) \right]$$

$$= \sum_k k P(k; \sigma) \int_0^T \frac{dt_1}{T} f(t - t_1) = \nu \int_0^T dt_1 f(t - t_1), \qquad (17.31)$$

where we have used the fact that the average of $k$ over the Poisson distribution gives $\sigma$, and $\sigma/T = \nu$. Suppose the width of $f(t)$ is $\Delta$. Then, as long as $T \gg \Delta$, the integral over $t_1$ is essentially the area under $f(t)$, except when $t$ lies within $\Delta$ of the endpoints of the time period. This is illustrated in Fig. 17.6. Since the latter cases are insignificant when $T \to \infty$, we have proved the first part of the theorem. The second part takes longer to prove, but the method is the same (see Problem 17.7).

Let $P(I)\,dI$ be the probability that the value of $I(t)$ lies between $I + dI$ and $I$. For large frequency $\nu$ of the events, $I(t)$ is a sum of a large number of stochastic variables, and therefore obeys a Gaussian distribution by the central limit theorem. Thus we have

$$P(I) = \frac{1}{\sqrt{2\pi b}} e^{-(I-a)^2/2b}, \qquad (17.32)$$

where

$$a = \langle I(t) \rangle,$$
$$b = \langle I^2(t) \rangle - \langle I(t) \rangle^2. \qquad (17.33)$$

## Problems

**17.1**   There were twelve rain showers last week, but none happened on Tuesday. Is it safe to leave the umbrella home this Tuesday?

**17.2**    An MIT student drives to school and parks illegally on the street. One week she received 12 parking tickets, all on either Monday or Wednesday. Would you advise her to park in a pay lot on Mondays and Wednesdays?

**17.3**    A man in Hong Kong gets off work at approximately 5 PM every day, and walks to the Mong Kok subway station, where he could take a northbound or southbound train from the same platform. The man takes the first train that comes. (Need we explain? He has a house in Lowu and a condo in Wanchai.) The trains run on a strict schedule: a northbound train leaves every five minutes, and a southbound train leaves every five minutes. Over the years he wound up north in Lowu 70% of the time. Why?

**17.4**    Describe a procedure to test the quality of a random-number generator, based on the Poisson distribution. Carry it out in your personal computer.

**17.5**    The behavior of a certain type of semiconductor diode can be modeled by the current–voltage characteristic

$$I = \begin{cases} 0 & (V < 0) \\ I_0\left[\exp\left(V/V_0\right) - 1\right] & (V \geq 0). \end{cases}$$

Find the probability density for the current in terms of the probability density for the voltage.

**17.6**    A device squares the input: $y = \alpha x^2$. Suppose the input $x$ has a Rayleigh probability density

$$P(x) = \begin{cases} (x/a)\exp\left(-x^2/2a\right) & (x \geq 0) \\ 0 & (x < 0) \end{cases}$$

Find the probability density for $y$.

**17.7**    The sound of raindrops is represented by the output function $I(t)=\sum_k f(t - t_k)$, as defined in (17.28). Follow the step outlined below to show that its correlation function is given by

$$G(\tau) \equiv \langle I(t)I(t + \tau)\rangle = \nu \int_{-\infty}^{\infty} dt\, f(t)f(t + \tau) + \left(\nu \int_{-\infty}^{\infty} dt\, f(t)\right)^2,$$

where $\nu$ is average frequency of raindrops. The symbol $\langle \cdots \rangle$ denotes an average over an ensemble of time periods of duration $T$, with fixed $t$ and $\tau$, in the limit $T \to \infty$. For $\tau = 0$, this is the second part of Campbell's theorem.

(a)    Consider first exactly $K$ raindrops falling during the period $T$; average over the times at which they fall, then average over a Poisson distribution $P_\sigma(K)$ of values of $K$. Show that

$$G(\tau) = \sum_{K=0}^{\infty} P_\sigma(K) \sum_{i=1}^{K} \sum_{j=1}^{K} \int_0^T \frac{dt_1}{T} \cdots \int_0^T \frac{dt_K}{T} f(t - t_i)f(t + \tau - t_j),$$

where $\sigma = \nu T$, and the times $T, t, \tau, t_i$ are all integers in appropriate units.

(b) Consider separately the contribution from terms with $i = j$, and those with $i \neq j$. Show that

$$G(\tau) = \sum_{K=0}^{\infty} P_\sigma(K) \left[ \frac{K}{T} \int_0^T dt_1 f(t - t_1) f(t + \tau - t_1) \right.$$

$$\left. + \frac{K(K-1)}{T^2} \int_0^T dt_1 f(t - t_1) \int_0^T dt_2 f(t + \tau - t_2) \right].$$

(c) Since $f(t)$ has a finite width, and $T \to \infty$, show that for all values of $t + \tau$ except for a negligible set within a width of the boundaries of $[0, T]$, we have

$$G(\tau) = \sum_{K=0}^{\infty} P_\sigma(K) \left\{ \frac{K}{T} \int_{-\infty}^{\infty} dt_1 f(t) f(t + \tau) \right.$$

$$\left. + \frac{K(K-1)}{T^2} \left[ \int_{-\infty}^{\infty} dt f(t) \right]^2 \right\}.$$

Obtain the final form by summing over $K$.

(d) For $f(t) = \theta(t) e^{-\lambda t}$, show that

$$G(\tau) = \left( \frac{\nu}{\lambda} \right)^2 + \left( \frac{\nu}{2\lambda} \right) e^{-\lambda |\tau|}.$$

# Chapter 18

# Time-series analysis

## 18.1 Ensemble of paths

A time-series is a stochastic variable $v(t)$ that depends on time. We have seen an example of this in the output of shot noise. Here, we study its time-dependent aspects in greater depth. (For reference see Wang 1945, Wax 1954.) For concreteness we can think of $v(t)$ as the velocity of a Brownian particle. It does not have a definite form, but is a member of an ensemble of paths, as illustrated in Fig.18.1. The various records of $v(t)$ describe the time evolution of identically constituted systems under the action of random forces in the environment.

As we can see in Fig. 18.1, there is a distribution of $v$-values at each time, but that does not describe the time-series completely. We need to specify the correlations

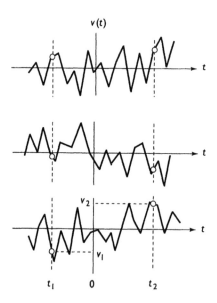

Figure 18.1 An ensemble of paths representing a stochastic process.

in time, and this is done by specifying a hierarchy of probability distributions. Using the abbreviation 1 for $\{v_1, t_1\}$, we list them as follows:

$$W_1(1),$$

$$W_2(1,2),$$

$$W_3(1,2,3)$$

$$\vdots \tag{18.1}$$

Here, $W_k(1,\ldots,k)\,dv_1,\ldots,dv_k$ is the joint probability of finding that $v$ has a value between $v_1$ and $v_1 + dv_1$ at time $t_1$, between $v_2$ and $v_2 + dv_2$ at time $t_2,\ldots$, and between $v_k$ and $v_k + dv_k$ at time $t_k$.

Clearly, $W_k$ must be positive-definite, and symmetric under the interchange of $\{v_i, t_i\}$ with $\{v_j, t_j\}$. The $n$th joint probability must imply all the lower ones $W_k$ with $k < n$:

$$W_k(1,\ldots,k) = \int_{-\infty}^{\infty} dv_{k+1} \ldots \int_{-\infty}^{\infty} dv_n \, W_n(1,\ldots,n). \tag{18.2}$$

In principle, we can measure these distributions from records like those in Fig. 18.1, given a sufficiently large sample. To find $W_2(1, 2)$, for example, we follow the dotted lines at $t_1$ and $t_2$ in Fig. 18.1, and find the fraction of records for which $v(t_1) = v_1$ and $v(t_2) = v_2$, to within specified tolerances.

We shall limit our attention to *stationary ensembles*, for which the probability distributions are invariant under time translation. This means that $W_1(v_1, t_1)$ is independent of $t_1$, and $W_k$ depends only on the relative times $t_1 - t_2$ etc.

Many physical processes approach a steady state after the transients have died down. In these cases, we admit only the time records taken during the steady state. In Brownian motion, the path of a single particle never reaches a steady state, but the velocity does.

For a stationary ensemble, time average is equivalent to ensemble average. The collection of time records, say from $t = 0$ to $t = T$, can be obtained from one time record, say the first one, by cutting out successive blocks of length $T$.

We can of course think of situations where time average is different from ensemble average, even in a stationary situation. Suppose we toss a coin $N$ times in succession. Is the number of heads the same as those obtained by tossing $N$ coins simultaneously as $N \to \infty$? The answer is obviously no, even in the large $N$ limit, for there is a possibility that in the first experiment we will get all heads. Such a sequence, however, is very unlikely for large $N$, and in the limit $N \to \infty$ they have measure zero. The equivalence between the time average and ensemble average is conditioned upon the neglect of such zero-measure cases.

## 18.2   Power spectrum and correlation function

There is an intuitive relation between spectrum and correlation, as G.I. Taylor explains so well (Taylor 1938):

> When a prism is set up in the path of a beam of white light it analyses the time variation of electric intensity at a point into its harmonic components and separates them into a spectrum. Since the velocity of light for all wavelengths is the same, the time variation analysis is exactly equivalent to a harmonic analysis of the space variation of electric intensity along the beam.

This idea is illustrated in Fig. 18.2. To formulate it mathematically, we begin with the Fourier expansion

$$v(t) = \int_{-\infty}^{\infty} \frac{d\omega}{2\pi} e^{-i\omega t} v_{\omega}, \tag{18.3}$$

with inverse transform

$$v_{\omega} = \int_{-\infty}^{\infty} dt \, e^{i\omega t} v(t). \tag{18.4}$$

Since $v(t)$ is real, we must have

$$v_{-\omega} = v_{\omega}^{*}. \tag{18.5}$$

The correlation function $\langle v(t_1)v(t_2) \rangle$ depends only on $t_1 - t_2$ for a stationary ensemble. Let us see what this implies for the Fourier transform:

$$\langle v(t_1)v(t_2) \rangle = \int_{-\infty}^{\infty} \frac{d\omega_1 \, d\omega_2}{(2\pi)^2} \langle v_{\omega_1} v_{\omega_2} \rangle \exp\left(-i\omega_1 t_1 - i\omega_2 t_2\right). \tag{18.6}$$

Putting $t_1 = T + \tau/2$, $t_2 = T - \tau/2$, we can write the exponent as

$$\omega_1 t_1 + \omega_2 t_2 = (\omega_1 + \omega_2)T + (\omega_1 - \omega_2)\frac{\tau}{2}. \tag{18.7}$$

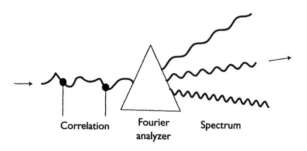

Correlation      Fourier       Spectrum
                 analyzer

Figure 18.2 Spectral analysis of a stream of events can reveal the correlations along the stream. For a stationary process, correlation at two different locations at the same time is equivalent to that at the same location at different times.

Thus

$$\langle v(t_1)v(t_2)\rangle = \int_{-\infty}^{\infty} \frac{d\omega_1 d\omega_2}{(2\pi)^2} \langle v_{\omega_1} v_{\omega_2}\rangle \exp\left(-i(\omega_1 + \omega_2)T - i(\omega_1 - \omega_2)\frac{\tau}{2}\right).$$
(18.8)

The right-hand side should be independent of $T$, and that means $\langle v_{\omega_1} v_{\omega_2}\rangle$ must be zero, unless $\omega_1 + \omega_2 = 0$. That is, it must have the form

$$\langle v_{\omega_1} v_{\omega_2}\rangle = 2\pi S(\omega_1)\delta(\omega_1 + \omega_2).$$
(18.9)

The coefficient $S(\omega)$ is called the *power spectrum*, which we assume to be real, with $S(\omega) = S(-\omega)$. When $S(\omega)$ is independent of $\omega$, we have *white noise*.
We can now write

$$\langle v(t_1)v(t_2)\rangle = \int_{-\infty}^{\infty} \frac{d\omega}{2\pi} S(\omega)e^{-i\omega(t_1 - t_2)}$$
(18.10)

or

$$\langle v(t)v(0)\rangle = \int_{-\infty}^{\infty} \frac{d\omega}{2\pi} S(\omega)e^{-i\omega t}$$

$$= \int_{0}^{\infty} \frac{d\omega}{\pi} S(\omega) \cos(\omega t).$$
(18.11)

The inverse Fourier transform gives

$$S(\omega) = \int_{-\infty}^{\infty} dt\, \langle v(t)v(0)\rangle e^{i\omega t}$$

$$= 2 \int_{0}^{\infty} dt\, \langle v(t)v(0)\rangle \cos(\omega t).$$
(18.12)

As expected on intuitive grounds, the power spectrum and the correlation function are Fourier transforms of each other. This relation, discovered by Wiener (Wiener 1930), and rediscovered by Kintchine (Kintchine 1934), is sometimes called the *Wiener–Kintchine theorem*. It appears to be trivial in our derivation, but not so to mathematicians who would not use the Dirac delta function.
Putting $t = 0$, we have

$$\langle v^2 \rangle = \int_{0}^{\infty} \frac{d\omega}{\pi} S(\omega).$$
(18.13)

If we think of $v(t)$ as a current, then the above equation represents the power dissipated in a unit resistance. The power dissipated in the frequency interval between $\omega$ and $\omega + d\omega$ is $\pi^{-1}S(\omega)$.

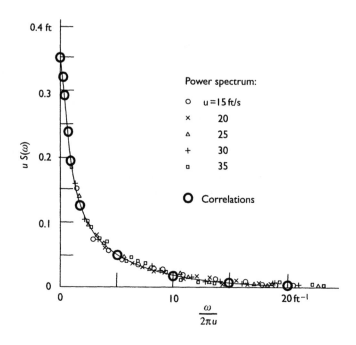

Figure 18.3 The power spectrum of turbulence in a wind tunnel, with average wind speed u. "Correlations" refer to the right-hand side of (18.14), which should be equal to the power spectrum by the Wiener–Kintchine theorem. After (Taylor 1938).

An experimental verification of the Wiener–Kintchine theorem was obtained by G.I. Taylor from measurements of local velocities $v$ in turbulence in a wind tunnel (Taylor 1938). To verify (18.12), $S(\omega)$ and $\langle v(t)v(0)\rangle$ should be measured at the same location in the wind tunnel. Equivalently, one can measure $\langle v(x/u)v(0)\rangle$ for different distances $x$ from the fixed location, where $u$ is the average velocity of the wind. Thus, the relation to be tested can be cast in the form

$$\frac{uS(\omega)}{\langle v^2\rangle} = \frac{2}{\langle v^2\rangle}\int_0^\infty dx \, \left\langle v\left(\frac{x}{u}\right)v(0)\right\rangle \cos\left(\frac{\omega x}{u}\right). \tag{18.14}$$

In Fig. 18.3, the two sides of the above equation are obtained through independent measurements, and plotted as functions of $\omega/u$, for a series of $u$'s. As we can see, they are in excellent agreement.

## 18.3  Signal and noise

From our perspective, noise is defined by the fact that all its Fourier components are stochastic variables with zero mean, i.e. $\langle v_\omega\rangle = 0$. A signal, therefore, is any definite additive periodic component.

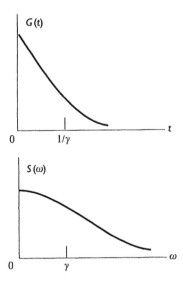

*Figure 18.4* The correlation function $G(t)$ and power spectrum $S(\omega)$ for a stochastic process without periodic components. White noise corresponds to the limit $\gamma \to \infty$.

If $v(t)$ does not contain periodic components, then the correlation function approaches zero when $t \to \infty$. As an illustrative example, let us assume that it decays exponentially:

$$G(t) = \langle v(t)v(0) \rangle = Ce^{-\gamma |t|}. \tag{18.15}$$

The corresponding power spectrum is easily found to be

$$S(\omega) = \frac{Cb}{\omega^2 + \gamma^2}. \tag{18.16}$$

This is known as a *Lorentzian distribution*. Qualitative plots of $G(t)$ and $S(\omega)$ are shown in Fig. 18.4. The power spectrum approaches white noise in the limit $\gamma \to \infty$.

Suppose there is a periodic component, so that $v(t)$ has the form

$$v(t) = u(t) + A \sin(\omega_0 t), \tag{18.17}$$

where $u(t)$ has no periodic component. Assuming

$$\langle u(t)u(0) \rangle = Ce^{-\gamma |t|}, \tag{18.18}$$

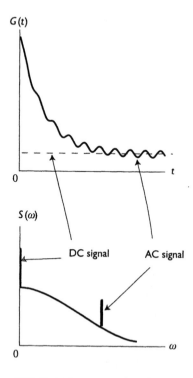

*Figure 18.5* Periodic components show up as spikes in the power spectrum.

we obtain

$$G(t) = Ce^{-\gamma|t|} + A^2 \sin(\omega_0 t),$$

$$S(\omega) = \frac{Cb}{\omega^2 + \gamma^2} + 2\pi A^2 \delta(\omega - \omega_0). \tag{18.19}$$

An ac signal of frequency $\omega_0$ will show up as a spike in the power spectrum at $\omega = \omega_0$. A dc signal will give a spike at $\omega = 0$. This is illustrated in Fig. 18.5.

## 18.4  Transition probabilities

We can describe the ensemble in terms of transition probabilities, which are conditioned probabilities that give the probability for an event when certain conditions are specified. We introduce the notation

$$P(1 \mid 2) = \text{Probability density of finding 2, when 1 is given}$$
$$P(1, 2 \mid 3) = \text{Probability density of finding 3, when 1,2 are given}$$
$$P(1, 2, 3 \mid 4) = \text{Probability density of finding 4, when 1, 2, 3 are given}$$

$$\vdots \qquad\qquad (18.20)$$

where 1 stands for $\{v_1, t_1\}$, etc, and probability density means probability per unit interval of $v$. The given variables are listed to the left of the vertical bar, and the variable to be found is to the right. They are related to the joint probability distributions as follows:

$$W_2(1, 2) = W_1(1)P(1 \mid 2)$$
$$W_3(1, 2, 3) = W_2(1, 2)P(1, 2 \mid 3)$$
$$W_3(1, 2, 3, 4) = W_2(1, 2, 3)P(1, 2, 3 \mid 4)$$

$$\vdots \qquad\qquad (18.21)$$

The transition probabilities must be positive-definite, and should satisfy the normalization condition

$$\int_{-\infty}^{\infty} dv_2 \, P(1 \mid 2) = 1. \qquad\qquad (18.22)$$

Since $W_1(2) = \int dv_1 W_2(1, 2)$, we have

$$W_1(2) = \int_{-\infty}^{\infty} dv_1 \, W_1(1)P(1 \mid 2). \qquad\qquad (18.23)$$

This is an integral equation for $W_1$ with $P(1 \mid 2)$ as the kernel.

Unlike $W_2(1, 2)$, $P(1 \mid 2)$ is not symmetric in $1, 2$. Using the symmetry of $W_2(1, 2)$, we obtain from (18.21) the relation

$$W_1(1)P(1 \mid 2) = W_1(2)P(2 \mid 1). \qquad\qquad (18.24)$$

That is, the transition probability weighted by the probability of the initial state is symmetric. This property is known as *detailed balance*.

## 18.5  Markov process

The simplest stochastic process is a "purely random" process, in which there are no correlations. All distributions are then determined by $W_1$ :

$$
\begin{aligned}
W_2(1,2) &= W_1(1)W_1(2), \\
W_3(1,2,3) &= W_1(1)W_1(2)W_1(3),
\end{aligned}
\tag{18.25}
$$

and so on. Such would be the case for the successive tossing of a coin. However, a continuous physical variable $v(t)$ cannot be purely random, because $v(t)$ and $v(t + dt)$ must be correlated, for sufficiently small $dt$.

Next in complexity comes the *Markov process*, which is defined by the property

$$
P(1, 2, \ldots, n-1 \,|\, n) = P(n-1 \,|\, n) \quad (t_n > t_{n-1} > \cdots > t_1). \tag{18.26}
$$

That is, the system has no memory beyond the last transition. All information about the process is contained in $P(1 \,|\, 2)$ or equivalently $W_2(1, 2)$. For example, from (18.21) we obtain

$$
\begin{aligned}
W_3(1, 2, 3) &= W_2(1, 2)P(12 \,|\, 3) = W_2(1, 2)P(2|3) \\
&= W_1(1)P(1 \,|\, 2)P(2 \,|\, 3).
\end{aligned}
\tag{18.27}
$$

The basic transition probability $P(1 \,|\, 2)$ cannot be arbitrary, but must satisfy the condition

$$
P(1|3) = \int_{-\infty}^{\infty} dv_2\, P(1 \,|\, 2)P(2 \,|\, 3). \tag{18.28}
$$

More explicitly,

$$
P(v_1, t_1 \,|\, v_3, t_3) = \int_{-\infty}^{\infty} dv_2 P(v_1, t_1 \,|\, v_2, t_2)P(v_2, t_2 \,|\, 3, t_3). \tag{18.29}
$$

This is called the *Smoluchowski equation*, or *Chapman–Kolmogorov equation*. Given that $P(i \,|\, j)$ are the only independent transition probabilities, this is obvious from the law of composition of probabilities implied by (18.2).

Markov processes are important, because most physical processes are of this type. Brownian motion falls into this category, and so do quantum mechanical transitions, where the transition probability per unit time is given by "Fermi's golden rule"

$$
\frac{\partial}{\partial t}P(1 \,|\, 2) = \frac{2\pi}{\hbar}|H'_{12}|^2 \rho_2, \tag{18.30}
$$

where $t = t_1 - t_2$, $H'_{12}$ is the matrix element of the interaction Hamiltonian, and $\rho_2$ is the density of final states.

## 18.6 Fokker–Planck equation

We now derive an equation for $P(1 \mid 2)$ in a Markov process, under the assumption that small displacements occur over small time intervals. First, let us make explicit the invariance under time translation by writing

$$P(1 \mid 2) = P(v_1, t_1 \mid v_2, t_2) \equiv P(v_1 \mid v_2, t_2 - t_1). \tag{18.31}$$

The Smoluchowski equation can be written in the form

$$P(v \mid u, t + \Delta t) = \int_{-\infty}^{\infty} dw \, P(v \mid w, t) P(w \mid u, \Delta t). \tag{18.32}$$

For a small time interval $\Delta t$, we assume that the following limits exist:

$$A(v) = \lim_{\Delta t \to 0} \frac{1}{\Delta t} \int_{-\infty}^{\infty} dw \, (w - v) P(v \mid w, \Delta t),$$

$$B(v) = \lim_{\Delta t \to 0} \frac{1}{\Delta t} \int_{-\infty}^{\infty} dw \, (w - v)^2 P(v \mid w, \Delta t). \tag{18.33}$$

The higher moments of $P(v \mid w, \Delta t)$ are assumed to be $O(\Delta t)^2$. These functions can be rewritten as

$$A(v) = \lim_{\Delta t \to 0} \frac{1}{\Delta t} \langle \Delta v \rangle,$$

$$B(v) = \lim_{\Delta t \to 0} \frac{1}{\Delta t} \langle (\Delta v)^2 \rangle, \tag{18.34}$$

where $v$ is an initial value that changes to $v + \Delta v$ in the time interval $\Delta t$, and we average $\Delta v$ over the ensemble.

Consider the integral

$$\int_{-\infty}^{\infty} du \, R(u) \frac{\partial P(v \mid u, t)}{\partial t} = \lim_{\Delta t \to 0} \frac{1}{\Delta t} \int_{-\infty}^{\infty} du \, R(u) \left[ P(v \mid u, t + \Delta t) - P(v \mid u, t) \right], \tag{18.35}$$

where $R(u)$ is an arbitrary function. Use the Smoluchowski equation to rewrite the first term on the right-hand side as

$$\int_{-\infty}^{\infty} du \, R(u) \int_{-\infty}^{\infty} dw \, P(v \mid w, t) P(w \mid u, \Delta t). \tag{18.36}$$

Now interchange the order of the integrations, and expand $R(u)$ about $u = w$ in a Taylor series:

$$R(u) = R(w) + (u - w) R'(w) + \tfrac{1}{2}(u - w)^2 R''(w) + \cdots \tag{18.37}$$

Substituting these results into (18.35), we obtain

$$\int_{-\infty}^{\infty} du\, R(u)\frac{\partial P(v|u,t)}{\partial t} = \int_{-\infty}^{\infty} dw\, P(v|w,t)\left[ R'(w)A(w) + \frac{1}{2}R''(w)B(w) \right].$$

(18.38)

On the right-hand side, we make partial integrations, and change the integration variable from $w$ to $u$, to rewrite it in the form

$$\int_{-\infty}^{\infty} du\left[ -R(u)\frac{\partial}{\partial u}(AP) + \frac{1}{2}R(u)\frac{\partial^2}{\partial u^2}(BP) \right],$$

(18.39)

where $A = A(u)$, $B = B(u)$, $P = P(v|u,t)$. Thus,

$$\int_{-\infty}^{\infty} du\, R(u)\left[ \frac{\partial}{\partial t}P + \frac{\partial}{\partial u}(AP) - \frac{1}{2}\frac{\partial^2}{\partial u^2}(BP) \right] = 0.$$

(18.40)

Since $R$ is arbitrary, we have the *Fokker–Planck equation*

$$\frac{\partial P}{\partial t} + \frac{\partial}{\partial u}(AP) - \frac{1}{2}\frac{\partial^2}{\partial u^2}(BP) = 0.$$

(18.41)

Here, $P = P(v|u,t)$ is a function of $u$ and $t$, with $v$ as the initial condition. The functions $A(u)$ and $B(u)$ depend on the dynamics of the system.

## 18.7 Langevin equation

To illustrate how we may incorporate dynamics into the discussion of a stochastic process, let $v$ be the velocity of a particle. The Newtonian equation of motion is

$$\frac{dv}{dt} + \gamma v = \frac{F}{m},$$

(18.42)

where $\gamma$ is a damping coefficient, and $F(t)$ is a randomly fluctuating external force. This is called the *Langevin equation*.

The same equation holds for the current in a RL circuit, if we identify $m = L$ and $m\gamma = R$, where $L$ is the inductance and $R$ is the resistance. In this case, $F$ is a fluctuating voltage due to thermal noise.

We specify the random force through the properties

$$\langle F_\omega \rangle = 0,$$
$$\langle F_\omega F_{\omega'} \rangle = 2\pi K(\omega)\delta(\omega + \omega'),$$
(18.43)

where $F_\omega = F^*_{-\omega}$ is the Fourier transform of $F(t)$. The correlation function associated with the force is

$$\langle F(t_1)F(t_2) \rangle = \int \frac{d\omega}{2\pi} K(\omega)e^{-i\omega t}.$$
(18.44)

For a white-noise spectrum, $K(\omega)$ is independent of $\omega$, and we have

$$\langle F(t_1)F(t_2) \rangle = 2\pi K_0 \delta(t_1 - t_2) \quad \text{(white noise)}.$$
(18.45)

This means that the correlation time is zero, i.e. much shorter than any other physical time scale in the problem. In Brownian motion, the correlation time for the force is of order $10^{-20}$ s, the molecular collision time. It is certainly much smaller than the observational time scale of seconds. Thus, for all practical purposes, a Brownian particle is driven by white noise.

Taking the Fourier transform of both sides of (18.42), we obtain

$$-i\omega v_\omega + \gamma v_\omega = \frac{F_\omega}{m},$$
(18.46)

$$v_\omega = \frac{F_\omega}{m} \frac{1}{\gamma - i\omega}.$$
(18.47)

The power spectrum of $v$ can be obtained by calculating

$$\langle v^*_\omega v_{\omega'} \rangle = \frac{1}{m^2} \frac{\langle F^*_\omega F_{\omega'} \rangle}{(\gamma + i\omega)(\gamma - i\omega')}$$

$$= \frac{2\pi K(\omega)}{m^2(\gamma^2 + \omega^2)}\delta(\omega - \omega').$$
(18.48)

Therefore the power spectrum of $v$ is

$$S(\omega) = \frac{K(\omega)}{m^2(\gamma^2 + \omega^2)}.$$
(18.49)

By the Wiener–Kintchine theorem, we have

$$\int_{-\infty}^{\infty} dt \, e^{-i\omega t} \langle v(t)v(0) \rangle = \frac{K(\omega)}{m^2(\gamma^2 + \omega^2)}.$$
(18.50)

In the Langevin equation, the force acting on a particle due to its environment is divided into a deterministic friction $\gamma v$ and a stochastic component $F$. The method works for linear systems. For nonlinear systems, the distinction between friction and random force can be ambiguous and can lead to inconsistencies (Van Kampen 1982).

## 18.8  Brownian motion revisited

We return to Brownian motion as an illustration of the more formal approach discussed in this chapter. We regard $v$ as the velocity, but our discussion applies equally well to Nyquist noise, or any system driven by white-noise force (18.45).

We begin by calculating the functions (18.34) for the Fokker–Planck equation. Integrating the Langevin equation (18.42) from a fixed initial value $v$, we have

$$\Delta v = -\gamma v \Delta t + \frac{1}{m} \int_t^{t+\Delta t} dt' F(t'). \tag{18.51}$$

It is necessary to integrate over $F$, even though $\Delta t$ is small, because $F$ fluctuates extremely rapidly. Using $\langle F \rangle = 0$, we obtain

$$A(v) = -\gamma v. \tag{18.52}$$

Next we need

$$\langle (\Delta v)^2 \rangle = \gamma^2 v^2 (\Delta t)^2 + \frac{1}{m^2} \int_t^{t+\Delta t} dt_1 \int_t^{t+\Delta t} dt_2 \, \langle F(t_1)F(t_2) \rangle . \tag{18.53}$$

Using (18.45), the double integral above can be shown to be $2\pi K_0 \Delta t$. Thus

$$B(v) = \frac{2\pi K_0}{m^2}. \tag{18.54}$$

The Fokker–Planck equation now reads

$$\frac{\partial P}{\partial t} - \gamma \frac{\partial}{\partial v}(vP) - \frac{\pi K_0}{m^2} \frac{\partial^2 P}{\partial v^2} = 0, \tag{18.55}$$

where $P(v,t)dv$ is the probability of finding $v$ within $dv$, at time $t$, when an initial value $v_0$ was specified at $t = 0$. The solution is

$$P(v,t) = \frac{1}{\sqrt{2\pi u}} e^{-(v-a)^2/2b}, \tag{18.56}$$

where

$$a = v_0 e^{-\gamma t},$$

$$b = \frac{K_0}{m^2 \gamma} \left(1 - e^{-\gamma t}\right). \tag{18.57}$$

This is a Gaussian distribution, as expected, since $v$ is the sum of a large number of Fourier components, which are stochastic variables. When $t \to \infty$, it approaches a stationary Gaussian with variance

$$\langle v^2 \rangle = \frac{K_0}{m^2 \gamma}. \tag{18.58}$$

This shows that the velocities of Brownian particles eventually attain a Maxwell–Boltzmann distribution.

We can obtain the coordinate distribution of Brownian particles from the velocity distribution. Chandrasekhar (Chandrasekhar 1943) has done this by a direct method. We follow an indirect but simpler route.

The coordinate $x$ must have a Gaussian distribution, because $v = \dot{x}$ does. Therefore all we have to do is find its variance. The Langevin equation is

$$\frac{d^2x}{dt^2} + \gamma \frac{dx}{dt} = \frac{F}{m}. \tag{18.59}$$

Multiply both sides by $x$ and take the ensemble average:

$$\left\langle x \frac{d^2x}{dt^2} \right\rangle + \gamma \left\langle x \frac{dx}{dt} \right\rangle = 0, \tag{18.60}$$

where we have used $\langle xF \rangle = 0$, because, while $x$ changes sign under reflection, $F$ does not. The above can be rewritten in the form

$$\frac{1}{2} \frac{d}{dt} \left\langle \frac{dx^2}{dt} \right\rangle - \langle v^2 \rangle + \frac{\gamma}{2} \left\langle \frac{dx^2}{dt} \right\rangle = 0, \tag{18.61}$$

where $\langle v^2 \rangle = u$. The solution for $t \to \infty$ is

$$\left\langle \frac{dx^2}{dt} \right\rangle = \frac{2u}{\gamma} = \frac{K_0}{m^2\gamma^2}, \tag{18.62}$$

or

$$\langle x^2 \rangle = \frac{K_0 t}{m^2\gamma^2}. \tag{18.63}$$

The diffusion constant is defined such that the above is $2Dt$. Therefore

$$D = \frac{K_0}{2m^2\gamma^2}. \tag{18.64}$$

Einstein's relation can be obtained by noting that $\langle v^2 \rangle = kT/m$ by the equipartition of energy. Comparison with (18.58) gives $K = m\gamma kT$, and this leads to

$$D = \frac{kT}{2m\gamma}. \tag{18.65}$$

Thus, the mobility is

$$\eta = \frac{1}{2m\gamma}. \tag{18.66}$$

Going back to (18.50) and putting $\omega = 0$, we obtain

$$\beta \int_0^\infty dt \langle v(t)v(0) \rangle = \eta, \tag{18.67}$$

where $\beta = 1/kT$. This is a more explicit form of the fluctuation–dissipation theorem.

## 18.9   The Monte-Carlo method

The Monte-Carlo method is a computer algorithm to simulate a thermal ensemble. An ideal statistical ensemble consists of an infinite number of copies of the system, for the purpose of taking thermal averages. No computer has a large enough memory for that. So we generate the members of the ensemble one at a time, and replace the ensemble average by a time average. The accuracy of a computation would be limited by computing time instead of computer memory.

Consider a system whose state is denoted by $C$, and the energy of the state by $E(C)$. In the canonical ensemble with temperature $T$, with $\beta = 1/kT$, the relative probability for the occurrence of $C$ in the ensemble is $e^{-\beta E(C)}$, and the thermodynamic average of any quantity $O(C)$ is given by

$$\langle O \rangle = \frac{\sum_C e^{-\beta E(C)} O(C)}{\sum_C e^{-\beta E(C)}}. \tag{18.68}$$

Our object is to instruct the computer to generate a sequence of states with the canonical distribution, i.e. states should be output with relative probability $e^{-\beta E(C)}$.

Let $f(C)$ be the probability distribution of a given ensemble. The equilibrium ensemble corresponds to

$$f_{eq}(C) = \frac{e^{-\beta E(C)}}{\sum_C e^{-\beta E(C)}}. \tag{18.69}$$

We want to generate a sequence of states $C_1 \rightarrow C_2 \rightarrow \cdots C_n + C_{n+1} \rightarrow \cdots$, which starts with an arbitrary initial state $C_1$, and, after a "warm up" period of $n$ steps, reaches a steady sequence of equilibrium states. From the $n$th step on, the "time average" with respect to the sequence should be equivalent to an ensemble average over a canonical ensemble.

The objective is achieved through a Markov process with transition probability $P(C_1|C_2)$ for $C_1 \rightarrow C_2$. It is the conditioned probability of finding the system in $C_2$, when it is initially in $C_1$. We impose the following conditions:

$$P(C_1 \,|\, C_2) \geq 0,$$

$$\sum_{C_2} P(C_1 \,|\, C_2) = 1,$$

$$e^{-\beta E(C_1)} P(C_1 \,|\, C_2) = e^{-\beta E(C_2)} P(C_2 \,|\, C_1). \tag{18.70}$$

The first two are necessary properties of any probability. The last is the statement of detailed balance, when the system is in contact with a heat reservoir.

**Theorem:** *The Markov process defined by (18.70) eventually leads to the equilibrium ensemble.*

*Proof*: Summing the detailed balance statement over states $C_1$, we obtain

$$\sum_{C_1} e^{-\beta E(C_1)} P(C_1 \mid C_2) = \sum_{C_1} e^{-\beta E(C_2)} P(C_2 \mid C_1). \tag{18.71}$$

On the right-hand side we note $\sum_{C_1} P(C_2 \mid C_1) = 1$. Thus

$$\sum_{C_1} e^{-\beta E(C_1)} P(C_1 \mid C_2) = e^{-\beta E(C_2)}. \tag{18.72}$$

This shows that the equilibrium distribution is an eigenstate of the transition matrix.
The "distance" between two ensembles $f_1(C)$ and $f_2(C)$ may be measured by

$$d(f_1, f_2) = \sum_C |f_1(C) - f_2(C)|. \tag{18.73}$$

Suppose $f_2$ is obtained from $f_1$ through a transition: $f_2(C) = \sum_{C'} f_1(C') P(C' \mid C)$.
The distance between $f_2$ and the equilibrium ensemble is given by

$$\sum_C |f_2(C) - f_{eq}(C)| = \sum_C \left| \sum_{C'} f_1(C') P(C' \mid C) - f_{eq}(C) \right|$$

$$= \sum_C \left| \sum_{C'} [f_1(C') - f_{eq}(C')] P(C' \mid C) \right|$$

$$\leq \sum_C \sum_{C'} |f_1(C') - f_{eq}(C')| P(C' \mid C), \tag{18.74}$$

where we have used (18.72) in the second step. Putting $\sum_C P(C' \mid C) = 1$ in the
last step, we obtain the inequality

$$\sum_C |f_2(C) - f_{eq}(C)| \leq \sum_C [f_1(C) - f_{eq}(C)]. \tag{18.75}$$

This shows that the distance from the equilibrium cannot decrease as the result of
a transition.                                                                      ■

The *Metropolis algorithm* gives a recipe in conformity with the rules (18.70),
as follows:

- Suppose the state is $C$.
- Make a trial change to $C'$.
- If $H(C') < H(C)$, accept the change.
- If $H(C') > H(C)$, accept the change conditionally, with probability
  $e^{-\beta[H(C')-H(C)]}$.

The conditional change in the last statement simulates thermal fluctuations. The relative transition probability corresponding to this algorithm is

$$T(C \mid C') = \begin{cases} 1 & \text{if } E(C') < E(C), \\ e^{-\beta[E(C')-E(C)]} & \text{if } E(C') > E(C). \end{cases} \tag{18.76}$$

The transition probability is obtained by properly normalizing the above:

$$P(C \mid C') = \frac{T(C \mid C')}{\sum_{C''} T(C \mid C'')}. \tag{18.77}$$

## 18.10   Simulation of the Ising model

We illustrate the Monte-Carlo method with the Ising model, which we discussed in Chapter 14. Consider a 2D square lattice. The $i$th lattice site has a spin $s_i = \pm 1$, with nearest-neighbor interaction energy $-\epsilon$. A state is specified by all the spins:

$$C = \{s_1, \ldots, s_N\}, \tag{18.78}$$

The energy of the state is

$$E(C) = -\epsilon \sum_{\langle ij \rangle} s_i s_j - h \sum_i s_i, \tag{18.79}$$

where $h$ is the external magnetic field, and where $\epsilon > 0$ for ferromagnetism, and $\epsilon < 0$ for antiferromagnetism, and $\langle ij \rangle$ denotes a nearest-neighbor pair of sites. We shall consider the ferromagnetic case. The magnetization is given by

$$M(C) = \sum_i s_i. \tag{18.80}$$

A nonzero ensemble average for $h = 0$ indicates spontaneous magnetization.

We set up the lattice in the computer by assigning memory locations to the sites, and choose a definite boundary condition. It is simplest to choose periodic boundary conditions. After initializing the lattice spins, say all spins up for a "cold" start, or random assignments for a "hot" start, we are ready to begin. We update the lattice by going through the spins one by one, and decide whether or not to flip it. This is called "one sweep" of the lattice. When examining $s_i$, we only need to know its four nearest-neighbor spins. The interaction energy of $s_i$ is given by

$$w_i = -\epsilon s_i \sum_{\text{nn } i} s_j - h s_i, \tag{18.81}$$

where nn stands for "nearest-neighbor to". If we flip $s_i$, then $w_i$ changes sign. The Metropolis algorithm says:

if $w_i > 0$,   flip $s_i$
if $w_i < 0$,   flip $s_i$ with probability $e^{-2\beta w_i}$. \quad (18.82)

After this is done, we go to the next spin and repeat the algorithm, until we have done this for all the spins in the lattice. This completes one sweep of the lattice, and produces a transition $C \rightarrow C'$. We keep making sweeps, until we think the lattice is "warmed up". Then we can start taking data.

For the Ising model, or any model in which the site variable $s$ has a limited range of possible values, there is a more efficient algorithm called the *heat bath method*, which amounts to "touching" the $i$th spin with a heat bath. The interaction energy of a spin $s$ is

$$w = s(\Sigma + h), \tag{18.83}$$

where $\Sigma$ is the sum of nearest-neighbor spins. We take the probability of flipping to be

$$\frac{e^{\beta w}}{e^{\beta w} + e^{-\beta w}}. \tag{18.84}$$

It is easily seen that this is a probability that satisfies detailed balance. Thus, if $s = 1$ the flip probability is

$$T = \frac{1}{1 + e^{-2\beta(\Sigma+h)}}. \tag{18.85}$$

If $s = -1$, the flip probability is

$$1 - T = \frac{1}{1 + e^{2\beta(\Sigma+h)}}. \tag{18.86}$$

The sum of nearest-neighbor spins has the possible values

$$\Sigma = 4, 2, 0, -2, -4. \tag{18.87}$$

We can calculate the values of $T$ and $1 - T$ for each of the values of $\Sigma$, and store them as a lookup table. This is done only once to initialize the program. During runtime, we decide whether or not to flip the spin $s$ by looking up $T$ if $s = 1$, and $1 - T$ if $s = -1$. It can be shown that one step in the heat bath method is equivalent to an infinite number of Metropolis steps. However, the heat bath method becomes unwieldy if $s$ has a large number of possible values.

Having given possible recipes for the transition probability, let us now describe the overall implementation of the simulation. We generate a sequence of states, with the first block of states regarded as warmups. The states after that in the sequence should have a canonical distribution, and are taken to be members of the equilibrium ensemble. The sequence as illustrated in the following scheme:

$$\underbrace{\left(C_1' \rightarrow \cdots \rightarrow C_K'\right)}_{\text{Warmups}} \rightarrow \underbrace{(C_1 \rightarrow C_2 \rightarrow \cdots)}_{\text{The ensemble}}. \tag{18.88}$$

A member of the ensemble $C_i$ is kept in memory only for as long as needed to perform measurements, and is overwritten by the next member. In the measurement process, we calculate the energy $E(C_i)$, the square of the energy $E^2(C_i)$, and the magnetization $M(C_i)$. These quantities are accumulates additively as $C_i$ is replaced by $C_{i+1}$, so that as the program proceeds, we keep running totals

$$E = E(C_1) + E(C_2) + \cdots$$
$$E^2 = E^2(C_1) + E^2(C_2) + \cdots \qquad (18.89)$$
$$M = M(C_1) + M(C_2) + \cdots$$

If desired, we can calculate and accumulate other quantities.

Suppose at the end of the run we generated a total of $K$ states in the ensemble. We then calculate the following ensemble averages

$$\langle E \rangle = \frac{1}{K} \sum_{i=1}^{K} E(C_i),$$

$$\langle E^2 \rangle = \frac{1}{K} \sum_{i=1}^{K} E^2(C_i), \qquad (18.90)$$

$$\langle M \rangle = \frac{1}{K} \sum_{i=1}^{K} M(C_i).$$

The heat capacity is given by

$$C = \frac{1}{kT^2} \left( \langle E^2 \rangle - \langle E \rangle^2 \right). \qquad (18.91)$$

These numbers constitute the output of the program in one complete run.

To calculate the statistical errors in the output, we have to make a large number of independent runs. The results for a measured quantity should have a Gaussian distribution, from which we can find the mean and variance.

## Problems

**18.1** The correlation function for "the sound of raindrops" is given in Problem 17.7.

(a) Show that the power spectrum is given by

$$S(\omega) = u \, |f_\omega|^2 + \langle I(t) \rangle^2 \, 2\pi \, \delta(\omega),$$

where $f_\omega$ is the Fourier transform of $f(t)$, the sound of a single raindrop, and $I = \sum_k f(t - t_k)$ is the output stream.

(b) Obtain $S(\omega)$ for $f(t) = \theta(t)e^{-\lambda t}$ and interpret the result. Is there a white-noise component?

**18.2** The shot noise in a diode is described by the voltage $V(t) = R \sum_k \varphi(t - t_k)$, where $\varphi(t)$ represents a pulse of current. The arrival times $t_k$ are random, with an average rate $\nu$. The output is put through a low-pass filter, so only low frequencies are observed. Find the power spectrum in the low-frequency limit .

**18.3** *Random telegraph signals*: A stream of random telegraph signals $I(t)$ in a time interval $[0, T]$ is illustrated in the accompanying figure, where $T \to \infty$ eventually. The signals have values either $a$ or $-a$, and are of random length. The zeros on the time axis are distributed according to a Poisson distribution, with average rate $\nu$. We consider an ensemble of time intervals. Find the correlation function and power spectrum, by going through the following steps.

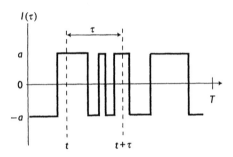

(a) The correlation function $\langle I(t)I(t + \tau)\rangle$ is obtained by averaging the product over the ensemble, with fixed $t$ and $\tau$. It is independent of $t$ because the ensemble is stationary. Referring to the illustration, we see that the product is $a^2$ if the factors have the same sign, and $-a^2$ if they have opposite signs. Thus show that

$$\langle I(t)I(t + \tau)\rangle = a^2 P_{\text{even}} - a^2 P_{\text{odd}},$$

where $P_{\text{even}}$, $P_{\text{odd}}$ are, respectively, the probability that there are an even and odd number of sign changes in the time interval $\tau$ .

(b) Show that

$$\langle I(t)I(t + \tau)\rangle = a^2 e^{-2\nu |\tau|}.$$

(c) Find the power spectrum.

**18.4** Show that the Smoluchowski equation (18.28) follows from the basic property of probabilities (18.2) and the definitions of transition probabilities (18.21). *Hint:* Start with the equation $W_3(3, 1, 2) = \int dx_4 W_4(3, 1, 4, 2)$ from (18.2), with the variables in the order shown. Express $W_3$ and $W_4$ in terms of $P(i \,|\, j)$ with the help of (18.21) and (18.26).

**18.5** *Probability of a path*: If we consider an ensemble of paths $x(t)$, we should be able to assign a probability $\mathcal{P}[x]$ that is a functional of the path. We do this in this problem for the case of Brownian motion.

(a) For Brownian motion, generalize the results in Section 16.5 to show that the transition probability density of finding $x$ at time $t$, given that it had the value

$x_0$ at time $t_0$, is

$$P(x, t \mid x_0, t_0) = \frac{1}{\sqrt{4\pi D(t - t_0)}} \exp\left(-\frac{(x - x_0)^2}{4D(t - t_0)}\right).$$

(b) Let a Brownian path $x(t)$ be specified by the positions $x_i$ at time $t_i$ ($i = 0, 1, \ldots, n$), with $t_0 < t_1 < \cdots < t_n$, as illustrated in the accompanying figure. The probability density for the path is given by

$$\mathcal{P}[x] = P(x_n, t_n \mid x_{n-1}, t_{n-1}) \cdots P(x_2, t_2 \mid x_1, t_1) P(x_1 t_1 \mid x_0, t_0).$$

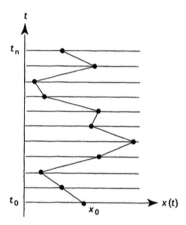

That is, when multiplied by $dx_0\, dx_1 \cdots dx_n$, this gives the probability that the Brownian particle is found between $x_i + dx_i$ and $x_i$, at time $t_i$, for $i = 0, 1, \ldots, n$. Show that it satisfies the law of composition of probabilities, i.e. if we sum this probability density over all possible paths having the same endpoints, we obtain the transition probability density to go from the initial point to the final point:

$$P(x_n, t_n \mid x_0, t_0) = \int_{-\infty}^{\infty} dx_{n-1} \cdots \int_{-\infty}^{\infty} dx_1\, \mathcal{P}[x].$$

*Hint*: Prove the relation for $n = 1$ and $n = 2$ by direct integration. Prove the general result by induction.

**18.6**  *Path-integral representation*: The probability for a Brownian path $x(t)$ can be represented by a Feynman path integral, as follows.

(a) Choose an equal time interval $\tau$ between the time steps. Using the explicit form of $P(x, t \mid x_0, t_0)$, show that

$$\mathcal{P}[x] = (4\pi D\tau)^{-n/2} \exp\left[-\frac{1}{4D} \sum_{i=0}^{n-1} \left(\frac{x_{n-1} - x_n}{\tau}\right)^2 \tau\right].$$

Take the limit $n \to \infty$, $\tau \to 0$ to obtain

$$P[x] = \mathcal{N} \exp\left(-\frac{1}{4D}\int_{t_0}^{t} dt'\dot{x}^2\right),$$

where $\dot{x}$ is the velocity of the path $x(t')$, and $\mathcal{N}$ is a normalization constant.

(b) Show that the transition probability can be written as the Feynman path integral:

$$P(x,t \,|\, x_0, t_0) = \mathcal{N} \int (dx) \exp\left(-\frac{1}{4D}\int_{t_0}^{t} dt'\dot{x}^2\right),$$

where the integral extends over all paths $x(t')$ with the conditions $x(t_0) = x_0$, $x(t) = x$. This result gives a path-integral representation of the solution to the diffusion equation.

**18.7**   Write and run a computer program to obtain the transition temperature $T_c$ of the 2D Ising model using the Monte-Carlo method. The exact value of $T_c$ is given by $\tanh^2(2\epsilon/kT_c) = 1/2$, or

$$\frac{kT_c}{\epsilon} = 2.269185.$$

# Appendix

# Mathematical reference

## A.1 Stirling's approximation

In statistical analyses we frequently encounter the factorial

$$n! = 1 \cdot 2 \cdots n. \tag{A.1}$$

For large $n$, Stirling's approximation gives

$$n! \approx n^n e^{-n} \sqrt{2\pi n}, \tag{A.2}$$

or

$$\ln n! \approx n \ln n - n + \ln \sqrt{2\pi n}. \tag{A.3}$$

Usually the first two terms suffice. The relative error of this formula is about $(12n)^{-1}$, which is about 2% for $n = 4$.

To derive Stirling's approximation, start with the representation

$$n! = \Gamma(n+1) = \int_0^\infty dt \, t^n e^{-t}. \tag{A.4}$$

The integrand has a maximum at $t = n$. The value at the maximum gives $n! \approx n^n e^{-n}$. Expanding the integrand about the maximum yields the corrections.

## A.2 The delta function

The Dirac delta function $\delta(x)$ is zero if $x \neq 0$, and infinite if $x = 0$, such that

$$\int dx \, \delta(x) = 1, \tag{A.5}$$

where the range of integration includes $x = 0$. It is not a function in the ordinary sense, since it's value is not defined where it is non-zero. We use it with the understanding that eventually it will find its way into an integral. Mathematicians call it a "distribution."

Some useful properties of the delta function are

$$\delta(x) = \delta(-x),$$

$$\delta(ax) = \frac{1}{|a|} \delta(x). \tag{A.6}$$

For any given function $f(x)$, we have

$$\int dx f(x) \delta(x - a) = f(a). \tag{A.7}$$

Its integral is the step function

$$\theta(x) = \begin{cases} 1 & (x > 0), \\ 0 & (x < 0). \end{cases} \tag{A.8}$$

The derivative of the delta function is the $\epsilon$-function

$$\epsilon(x) = \begin{cases} 1 & (x > 0), \\ -1 & (x < 0). \end{cases} \tag{A.9}$$

The Fourier analysis is given by

$$\delta(x) = \int \frac{dk}{2\pi} e^{ikx}. \tag{A.10}$$

In higher dimensions the delta function is defined as the product of 1D delta functions. For example,

$$\delta^3(\mathbf{r}) = \delta(x)\delta(y)\delta(z) = \int \frac{d^3k}{(2\pi)^3} e^{i\mathbf{k}\cdot\mathbf{x}}. \tag{A.11}$$

## A.3  Exact differential

The differential

$$df = A(x, y)\, dx + B(x, y)\, dy \tag{A.12}$$

is said to be an "exact differential" if there exists a function $f(x, y)$ which changes according to the above when its independent variables are changed. We must have

$$A(x, y) = \frac{\partial f}{\partial x},$$

$$B(x, y) = \frac{\partial f}{\partial y}. \tag{A.13}$$

Since differentiation is commutative, we have

$$\frac{\partial A}{\partial y} = \frac{\partial B}{\partial x}. \tag{A.14}$$

This is the condition for an exact differential.

## A.4  Partial derivatives

Suppose three real variables $x$, $y$ and $z$ are constrained by one condition $f(x, y, z) = 0$, where $f$ is a regular function. Partial derivatives have the properties

$$\left(\frac{\partial x}{\partial y}\right)_w \left(\frac{\partial y}{\partial z}\right)_w = \left(\frac{\partial x}{\partial z}\right)_w,$$

$$\left(\frac{\partial x}{\partial y}\right)_w = \frac{1}{\left(\frac{\partial y}{\partial x}\right)_w}, \tag{A.15}$$

where $w$, the quantity being held fixed, is some function of the variables.

## A.5  Chain rule

The *chain rule* states

$$\left(\frac{\partial x}{\partial y}\right)_z \left(\frac{\partial y}{\partial z}\right)_x \left(\frac{\partial z}{\partial x}\right)_y = -1 \tag{A.16}$$

This follows from

$$df = \frac{\partial f}{\partial x}\,dx + \frac{\partial f}{\partial y}\,dy + \frac{\partial f}{\partial z}\,dz = 0, \tag{A.17}$$

where $\partial f/\partial x$ denotes the partial derivative with respect to $x$, with the other two variables kept fixed. From this we get

$$\left(\frac{\partial x}{\partial y}\right)_z = -\frac{\partial f/\partial y}{\partial f/\partial x},$$

$$\left(\frac{\partial y}{\partial z}\right)_x = -\frac{\partial f/\partial z}{\partial f/\partial y}, \tag{A.18}$$

$$\left(\frac{\partial z}{\partial x}\right)_y = -\frac{\partial f/\partial x}{\partial f/\partial z}.$$

The desired relation is obtained by multiplying the above together.

## A.6  Lagrange multipliers

Consider a function of $n$ variables $f(\mathbf{x})$, where $\mathbf{x}$ represents an $n$-component vector. An extremum of $f(\mathbf{x})$ is determined by

$$df(\mathbf{x}) = d\mathbf{x} \cdot \nabla f(\mathbf{x}) = 0, \qquad (A.19)$$

where $d\mathbf{x}$ represents an arbitrary infinitesimal change in $x$. If, however, there exists a constraint

$$g(\mathbf{x}) = 0, \qquad (A.20)$$

then $dx$ cannot be completely arbitrary, but must be such as to maintain $dg = 0$ or

$$dg = d\mathbf{x} \cdot \nabla g = 0. \qquad (A.21)$$

This requires $dx$ to be tangent to the surface $g = 0$, as represented schematically in Fig. A.1. We have to find the extremum of $f(\mathbf{x})$ on the surface $g(\mathbf{x}) = 0$.

An arbitrary differential $d\mathbf{x}$ can be resolved into transverse (orthogonal) and longitudinal (tangential) components with respect to the given surface:

$$d\mathbf{x} = d\mathbf{x}_T + d\mathbf{x}_L. \qquad (A.22)$$

The condition for an extremum is

$$d\mathbf{x}_L \cdot \nabla f = 0. \qquad (A.23)$$

Since $d\mathbf{x}_L \cdot \nabla g = 0$ by definition, the above can be generalized to

$$d\mathbf{x}_L \cdot \nabla (f + \lambda g) = 0, \qquad (A.24)$$

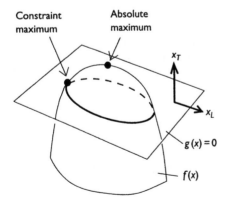

Figure A.1  Illustrating the method of Lagrange multipliers to find a constraint maximum.

where $\lambda$ is an arbitrary number. Let us choose $\lambda$ such that it satisfies

$$d\mathbf{x}_T \cdot \nabla(f + \lambda g) = 0. \tag{A.25}$$

Then, by adding this equation to the previous one, we have $d\mathbf{x} \cdot \nabla(f + \lambda g) = 0$, or

$$d[f(\mathbf{x}) + \lambda g(\mathbf{x})] = 0, \tag{A.26}$$

without any constraint. The parameter $\lambda$ is called a Lagrange multiplier. To solve the problem, we can first find $\mathbf{x}$ for arbitrary $\lambda$. Denoting the result by $\mathbf{x}(\lambda)$, we determine $\lambda$ by requiring $g(\mathbf{x}(\lambda)) = 0$. This determines the location of the desired extremum $\mathbf{x}(\lambda)$.

To summarize: the extremum of $f$ subject to the constraint $g = 0$ can be found by finding the extremum of $f + \lambda g$, and then determining $\lambda$ so as to satisfy the constraint.

## A.7  Counting quantum states

A free particle in quantum mechanics is described by the wave function

$$\phi_\mathbf{k}(\mathbf{r}) = \frac{1}{\sqrt{V}} e^{i\mathbf{k}\cdot\mathbf{r}}, \tag{A.27}$$

where the wave vector $\mathbf{k}$ labels the state. This is normalized to one particle in the volume $V$. Periodic boundary conditions require

$$e^{i\mathbf{k}\cdot\mathbf{r}} = e^{i\mathbf{k}\cdot\mathbf{r}+i\mathbf{k}\cdot\mathbf{m}L}, \tag{A.28}$$

where $\mathbf{m}$ is a vector whose components have possible values $0, \pm 1, \pm 2, \ldots$. Thus $\mathbf{k} \cdot \mathbf{m}L = 0 \bmod(2\pi)$, and the spectrum of $\mathbf{k}$ is given by:

$$\mathbf{k} = \frac{2\pi \mathbf{n}}{L}, \tag{A.29}$$

where $\mathbf{n}$ is a vector whose components have possible values $0, \pm 1, \pm 2, \ldots$. Since the spacing between successive values is $2\pi/L$, the spectrum approaches a continuum as $L \to \infty$.

A sum over states is denoted by:

$$\sum_\mathbf{k} = \sum_{n_1=-\infty}^{\infty} \sum_{n_2=-\infty}^{\infty} \sum_{n_3=-\infty}^{\infty}. \tag{A.30}$$

In the limit $L \to \infty$, this approaches a multiple integral. The spacing between values of $k_1$ is given by

$$\Delta k_1 = \frac{2\pi}{L} \Delta n_1, \tag{A.31}$$

and $\Delta n_1 = 1$. Thus

$$\sum_{n_1=-\infty}^{\infty} = \sum_{n_1=-\infty}^{\infty} \Delta n_1 = \frac{L}{2\pi} \sum_{k_1=-\infty}^{\infty} \Delta k_1 \rightarrow \frac{L}{2\pi} \int_{-\infty}^{\infty} dk_1. \tag{A.32}$$

Therefore

$$\sum_{k} \rightarrow \frac{V}{(2\pi)^3} \int d^3k \tag{A.33}$$

In terms of the momentum $\mathbf{p} = \hbar \mathbf{k}$, we have

$$\sum_{p} \rightarrow \frac{V}{(2\pi\hbar)^3} \int d^3p = \int \frac{d^3r\, d^3p}{h^3}. \tag{A.34}$$

This shows that the volume of a basic cell in phase space is

$$\tau_0 = h^3. \tag{A.35}$$

## A.8   Fermi functions

The Fermi functions have the power series expansions

$$f_n(z) \equiv \sum_{\ell=1}^{\infty} (-1)^{\ell+1} \frac{z^\ell}{\ell^n}. \tag{A.36}$$

We illustrate the large $z$ behavior by calculating that for $f_{3/2}(z)$. Go back to the integral representation (9.xx), and put $y = x^2$, $z = e^\nu$ :

$$f_{3/2}(z) = \frac{2}{\sqrt{\pi}} \int_0^\infty dy \frac{\sqrt{y}}{e^{y-\nu}+1}. \tag{A.37}$$

For $\nu$, the factor $(e^{y-\nu}+1)^{-1}$, which is the occupation number, is nearly a step function, whose derivative is nearly a delta function. If we can rework the integrand into something involving that derivative, then most of the contribution to the integral would come from the neighborhood of the Fermi surface $y = \nu$. With this goal in mind, we make a partial integration:

$$f_{3/2}(z) = \frac{2}{\sqrt{\pi}} \left\{ \frac{2}{3} \left[ \frac{y^{3/2}}{e^{y-\nu}+1} \right]_0^\infty - \frac{2}{3} \int_0^\infty dy\, y^{3/2} \frac{\partial}{\partial y} \frac{1}{e^{y-\nu}+1} \right\}$$

$$= \frac{4}{3\sqrt{\pi}} \int_0^\infty dy\, y^{3/2} \frac{e^{y-\nu}}{(e^{y-\nu}+1)^2}. \tag{A.38}$$

The integrand is now peaked at $y = v$. We put $y = v + t$ and obtain

$$f_{3/2}(z) = \frac{4v^{3/2}}{3\sqrt{\pi}} \int_{-v}^{\infty} dt \left(1 + \frac{t}{v}\right)^{3/2} \frac{e^t}{(e^t + 1)^2}. \tag{A.39}$$

We are interested in large $v$, so the lower limit is near $-\infty$. Accordingly, we write

$$f_{3/2}(z) = \frac{4v^{3/2}}{3\sqrt{\pi}} \int_{-\infty}^{\infty} dt \left(1 + \frac{t}{v}\right)^{3/2} \frac{e^t}{(e^t + 1)^2} + O(e^{-v}). \tag{A.40}$$

The asymptotic expansion is obtained by neglecting $O(e^{-v})$, and expanding the factor $(1 + t/v)^{3/2}$ in inverse powers of $v = \ln z$:

$$f_{3/2}(z) \approx \frac{4v^{3/2}}{3\sqrt{\pi}} \int_{-\infty}^{\infty} dt \left(1 + \frac{3}{2}\frac{t}{v} + \frac{3}{8}\frac{t^2}{v^2} + \cdots\right) \frac{e^t}{(e^t + 1)^2}. \tag{A.41}$$

In the power series, only terms of even power in $t$ survive the integration. Thus

$$f_{3/2}(z) \approx \frac{4v^{3/2}}{3\sqrt{\pi}} \left(I_0 + \frac{3}{8v^2}I_2 + \cdots\right), \tag{A.42}$$

where

$$I_n = 2 \int_0^{\infty} dt \frac{t^n e^t}{(e^t + 1)^2},$$

$$I_0 = 1, \tag{A.43}$$

$$I_2 = \frac{\pi^2}{3}.$$

It is interesting to note

$$I_n = (n - 1)!(2n)(1 - 2^{1-n})\zeta(n) \quad (n \text{ even}), \tag{A.44}$$

where $\zeta(z) = \sum_{\ell=1}^{\infty} \ell^{-z}$ is the Riemann zeta function, a celebrated function of number theory.

For our purpose we only need the first few terms in the asymptotic expansion:

$$f_{3/2}(z) \approx \frac{4}{3\sqrt{\pi}} \left[(\ln z)^{3/2} + \frac{\pi^2}{8} \frac{1}{\sqrt{\ln z}} + \cdots\right]. \tag{A.45}$$

# Notes

## 16 Noise

1 Carnot's principle is what we call the second law of thermodynamics.
2 'Mastic' is an ingredient used in the preparation of varnish, from the bark of *Pistacia lentiscus* from Chios Island.

## 17 Stochastic processes

1 From the *Oxford English Dictionary*: Stochastic, a. Now *rare* or *obs*. Pertaining to conjecture. **1720** Swift, *Right of Preced. betw. Physicians & Civilians 11*, I am Master of the Stochastik Art, and by Virtue of that, I divine, that those Greek Words have crept from the Margin into the Text. **1688** Cudworth, *Freewill (1838) 40* There is need and use of this stochastical judging and opining concerning truth and falsehood in human life.

# References

Chandrasekhar, S. *Rev. Mod. Phys.*, **15**, 1 (1943), see eq. (172).

De Gennes, P.G. *Superconductivity of Metals and Alloys*. W.A. Benjamin, New York (1966).

deShalit, A. and Feshbach, H. *Theoretical Nuclear Physics*, Vol. 1, p. 11. John Wiley, New York (1974).

Donelly, R.J. *et al. J. Low Temp. Phys.*, **44**, 471 (1981).

Einstein, A. *Ann. d. Phys.*, **17**, 549 (1905); **19**, 371 (1906).

Feller, W. *An Introduction to Probability Theory and its Applications*, 3rd edn. John Wiley, New York (1968).

Glyde, H.R. *et al. Europhys. Lett.*, **43**, 422 (1998).

Guggenheim, E.A. *J. Chem. Phys.*, **13**, 253 (1945).

Guth, A. In *The Oskar Klein Memorial Lectures* (ed. G. Ekspong), Vol. 2. World Scientific Publishing, Singapore (1992).

Holborn, L. and Otto, K. *Z. Physik*, **33**, 1 (1925).

Hill, R.W. and Lounasmaa, O.V. *Phil. Mag.*, **2**, 143 (1957).

Hanbury-Brown, R. and Twiss, R.O. *Nature*, **177**, 27 (1957).

Huang, K. *Statistical Mechanics*, 2nd edn. John Wiley, New York (1987).

Huang, K. *Quarks, Leptons, and Gauge Fields*, 2nd edn. World Scientific Publishing, Singapore (1992).

Huang, K. *Quantum Field Theory, from Operators to Path Integrals*. John Wiley, New York (1998).

Johnson, J.B. *Phys. Rev.*, **32**, 97 (1928).

Keesom, P.H. and Pearlman, N. *Phys. Rev.*, **91**, 1354 (1953).

Ketterle, W., Durfee, D.S. and Stamper-Kurn, D.M. In *Bose–Einstein Condensation in Atomic Gases, Proceedings of the International School of Physics "Enrico Fermi"*, Course CXL (eds. M. Inguscio, S. Stringari and C.E. Wieman) pp. 67–176. IOS Press, Amsterdam (1999).

Kintchine, A. *Math. Ann.*, **109**, 604 (1934).

Leff, H.S. *Maxwell's Demon: Entropy, Information, Computing* (eds. H.S. Leff and A.F. Rex). Princeton University Press, Princeton (1990).

Mather, J.C. *et al. Ap. J.*, **354**, L37 (1990).

Mewes, M.-O. *et al. Phys. Rev. Lett.*, **77**, 416 (1996).

Nyquist, H. *Phys. Rev.*, **32**, 110 (1928).

Perrin, M.J. *Ann. Chem. et Phys.*, 8me series (September, 1909) (translated from the French by F. Soddy), reprinted in M.J. Nye, *The Question of the Atom: From the Karlsruhe Congress to the Solvay Conference*, 1860–1911, Tomash Publishers, Los Angeles (1984).

Present, R.D. *Kinetic Theory of Gases*. McGraw-Hill, New York (1958).

Purcell, E. *Nature*, **178**, 1449 (1956).

Stenger, J. *et.al. J. Low Temp. Phys.*, **113**, 167 (1998).

Szilard, L. *Z. f. Physik*, **53**, 840 (1929).

Taylor, G.I. *Proc. Roy. Soc.*, **A** 164 (1938), included in *The Scientific Papers of Sir Geoffrey Ingram Taylor*, Vol. II (ed. G.K. Batchelor). Cambridge University Press, Cambridge, England (1960).

Thom, R. *Structural Stability and Morphogenesis*. Benjamin, New York, in French 1972, English translation by D.H. Fowler 1975.

Van Kampen, N.G. *Stochastic Processes in Physics and Chemistry*, p. 246. North-Holland, Amsterdam (1981).

Walker, C.B. *Phys. Rev.*, **103**, 547 (1956).

Wang, M.C. and Uhlenbeck, G.E. *Rev. Mod. Phys.*, **17**, 113 (1945).

Wannier, G.H. *Statistical Physics*. John Wiley, New York (1966).

Wax, N. (ed.). *Selected Papers on Noise and Stochastic Processes*. Dover Publications, New York (1954).

Wiener, N. *Acta Math.*, **55**, 117 (1930).

Wilson, A.H. *Thermodynamics and Statistical Mechanics*. Cambridge University Press, Cambridge (1957).

Zeeman, E.C. (ed.) *Catastrophe Theory, Selected Papers 1972–1977*. Addison-Wesley, Reading, MA (1977).

# Index

# Physical constants

| | |
|---|---|
| Planck's constant | $h = 6.626 \times 10^{-27}$ erg s |
| | $h = h/2\pi = 1.054 \times 10^{-27}$ erg s |
| Velocity of light | $c = 3 \times 10^{10}$ cm s$^{-1}$ |
| Proton charge | $e = 4.803 \times 10^{-10}$ esu $= 1.602 \times 10^{-19}$ C |
| Proton mass | $M_p = 1.673 \times 10^{-24}$ g |
| Electron mass | $m_e = 9.110 \times 10^{-28}$ g |
| Avogadro's number | $A_0 = 6.022 \times 10^{23}$ |
| Boltzman's constant | $k = 1.381 \times 10^{-16}$ erg K$^{-1}$ |
| Gas constant | $R = 8.314 \times 10^7$ erg K$^{-1}$mol$^{-1}$ |
| Density of water at STP | $1$ g cm$^{-3}$ |
| Specific heat capacity ($C_P$) of water at STP | $1$ cal g$^{-1}$K$^{-1}$ |

# Conversions

## Length, weight, temperature

$1$ litre $= 10^3$ cm$^3$
$1$ in $= 2.54$ cm
$1$ lb $= 0.453$ kg
$X°C = (273.15 + X)$ K
$X°F = \frac{5}{9}(X - 32)°C$

## Energy

$1$ J $= 10^7$ erg
$1$ eV $= 1.602 \times 10^{-12}$ erg
$1$ eV/$k = 1.160 \times 10^4$ K
$1$ cal $= 4.184$ J
$1$ Btu $= 1055$ J
$1$ ft lb $= 1.356$ J
$1$ kwh $= 3.6 \times 10^6$ J

## Power

$1$ W $= 1$ J s$^{-1}$
$1$ hp $= 550$ ft lb s$^{-1}$ $= 746$ W

## Pressure

$1$ atm $= 1.013 \times 10^6$ dyn cm$^{-2}$ $= 760$ mm Hg
$1$ bar $= 750$ mg Hg $= 10^6$ dyn cm$^{-2}$
$1$ torr $= 1$ mm Hg $= 1333$ dyn cm$^{-2}$
$1$ Pa $= 10^{-5}$ bar $= 10$ dyn cm$^{-2}$

Milton Keynes UK
Ingram Content Group UK Ltd.
UKHW021618071024
449327UK00020BA/1092

9 780748 409426